Family, Children, and Tourism in China

This edited volume explores various issues in family tourism studies and complements the dramatic development of this market segment in China. The book concentrates on family and children tourism, and through its chapters, hopes to enrich the landscape of family tourism in academia.

The family market in tourism has received increasing attention over past decades. Yet academic endeavors in this area remain somewhat lacking in depth and scope. In addition to imbalanced contributions from authors of diverse backgrounds, the extant literature suffers from insufficient inclusion of children. Relevant studies are largely limited to conventional tourism destinations such as beaches and cultural attractions. In response to growing academic interest in family tourism, this book is a compilation of eight chapters that attempt to push the scope and boundaries of existing research on family tourism.

The chapters in this book were originally published as a special issue of the *Journal of China Tourism Research.*

Mimi Li is Associate Professor in the School of Hotel and Tourism Management at The Hong Kong Polytechnic University. Dr Li is specialized in tourism marketing, tourist behavior, tourism planning, and China tourism-related issues.

Xinran Y. Lehto is Professor of Hospitality and Tourism Management at Purdue University. Dr Lehto is specialized in destination experience management and marketing consumer experiences and wellness in tourism issues.

Family, Children, and Tourism in China

Edited by
Mimi Li and Xinran Y. Lehto

Routledge
Taylor & Francis Group
LONDON AND NEW YORK

First published 2022
by Routledge
2 Park Square, Milton Park, Abingdon, Oxon OX14 4RN

and by Routledge
605 Third Avenue, New York, NY 10158

Routledge is an imprint of the Taylor & Francis Group, an informa business

British Library Cataloguing in Publication Data
A catalogue record for this book is available from the British Library

ISBN: 978-1-032-15396-4 (hbk)
ISBN: 978-1-032-15397-1 (pbk)
ISBN: 978-1-003-24398-4 (ebk)

DOI: 10.4324/9781003243984

Typeset in Minion Pro
by Newgen Publishing UK

Publisher's Note
The publisher accepts responsibility for any inconsistencies that may have arisen during the conversion of this book from journal articles to book chapters, namely the inclusion of journal terminology.

Contents

Citation Information

The chapters in this book were originally published in the *Journal of China Tourism Research*, volume 16, issue 1 (2020). When citing this material, please use the original page numbering for each article, as follows:

Introduction

40 Years of Family Tourism Research: Bibliometric Analysis and Remaining Issues
Mimi Li, Xinran Lehto and Huahua Li
Journal of China Tourism Research, volume 16, issue 1 (2020), pp. 1–22

Chapter 1

The Grand Tour in the Twenty-first Century: Perspectives of Chinese Adolescents and Their Parents
Zhuowei (Joy) Huang and Qian Li
Journal of China Tourism Research, volume 16, issue 1 (2020), pp. 23–44

Chapter 2

Exploring the Influence of Family Holiday Travel on the Subjective Well-being of Chinese Adolescents
Mingjie Gao, Mark E. Havitz and Luke R. Potwarka
Journal of China Tourism Research, volume 16, issue 1 (2020), pp. 45–61

Chapter 3

The Constructions of Family Holidays in Young Middle-class Malaysian Chinese Children
Catheryn Khoo-Lattimore and Mona Jihyun Yang
Journal of China Tourism Research, volume 16, issue 1 (2020), pp. 62–77

Chapter 4

Spatial Memory Bias in Children Tourists
Xiaoting Huang, Linlin Zhang and Lucie Ihnatoliova
Journal of China Tourism Research, volume 16, issue 1 (2020), pp. 78–95

Chapter 5

What do Parents Look for in an Overseas Youth Summer Camp? Perspectives of Chinese Parents
Xinran Y. Lehto, Xiaoxiao Fu, Ksenia Kirillova and Chen Bi
Journal of China Tourism Research, volume 16, issue 1 (2020), pp. 96–117

Chapter 6

Vacation Travel, Marital Satisfaction, and Subjective Wellbeing: A Chinese Perspective
Liping Cai, Saerom Wang and Yunzi Zhang
Journal of China Tourism Research, volume 16, issue 1 (2020), pp. 118–139

Chapter 7

Is Airbnb a Good Choice for Family Travel?
Pearl M. C. Lin
Journal of China Tourism Research, volume 16, issue 1 (2020), pp. 140–157

Notes on Contributors

Chen Bi is Sales Specialist at the Westin Beijing Chaoyang, Beijing, China.

Liping Cai is Professor and Director in the School of Hospitality and Tourism Management at Purdue University, USA.

Xiaoxiao Fu is Assistant Professor in the Rosen College of Hospitality Management at the University of Central Florida, Orlando, Florida, USA.

Mingjie Gao is PhD candidate of the Department of Recreation and Leisure Studies at the University of Waterloo. Her research interests include family vacations, travel and health, and coastal governance.

Mark E. Havitz is Professor of the Department of Recreation and Leisure Studies at the University of Waterloo. His teaching and research interests focus on consumer behavior, marketing, finance, and management.

Xiaoting Huang is Professor in the School of Management at Shandong University, Shandong, China. The author's research interests are tourism behavior and experiment method in tourism Research.

Zhuowei (Joy) Huang is Assistant Professor in the Department of Recreation, Sport and Tourism at the University of Illinois at Urbana-Champaign, Illinois, USA.

Lucie Ihnatoliova is Doctoral Student in the School of Management at Shandong University, Shandong, China. The author's research interests are tourism behavior and experiment method in tourism research.

Catheryn Khoo-Lattimore is Senior Lecturer in the Department of Tourism, Sport and Hotel Management, Griffith Business School at Griffith University, Queensland, Australia.

Ksenia Kirillova is Assistant Professor in the School of Hotel and Tourism Management at The Hong Kong Polytechnic University, Hong Kong SAR, China.

Xinran Y. Lehto is Professor of Hospitality and Tourism Management at Purdue University. Dr. Lehto is specialized in destination experience management and marketing consumer experiences and wellness in tourism issues.

Huahua Li School of Hotel and Tourism Management, the Hong Kong Polytechnic University, Hong Kong SAR, China. She is a PhD student at Nankai University.

Mimi Li is Associate Professor in the School of Hotel and Tourism Management at The Hong Kong Polytechnic University. Dr Li is specialized in tourism marketing, tourist behavior, tourism planning, and China tourism-related issues.

Qian Li is Doctoral Student of the Department of Recreation, Sport and Tourism at University of Illinois at Urbana-Champaign, Illinois, USA.

Pearl M. C. Lin is Assistant Professor in the School of Hotel and Tourism Management at the Hong Kong Polytechnic University. She is interested in food and beverage management, as well as sharing economy in accommodation, dining, and food delivery services.

Luke R. Potwarka is Assistant Professor of the Department of Recreation and Leisure Studies at the University of Waterloo. His research focuses on consumer behavior related to sport events.

Saerom Wang is affiliated with the School of Hospitality and Tourism Management at Purdue University, USA.

Mona Jihyun Yang is a PhD student at Griffith University.

Linlin Zhang is Doctoral Student in the School of Management at Shandong University, Shandong, China. The author's research interests are tourism behavior and experiment method in tourism research.

Yunzi Zhang is Associate Professor in the Business Department at Northern Marianas College, Saipan, Commonwealth of the Northern Mariana Islands.

Introduction: 40 Years of Family Tourism Research-Bibliometric Analysis and Remaining Issues

Mimi Li, Xinran Y. Lehto and Huahua Li

ABSTRACT

Based on a bibliometric analysis of research on family tourism, this editorial intends to reflect on the knowledge development in the topic area in the past 40 years. Eight themes or subject clusters were unveiled: foundations, family roles, place, family structure and function, psychological factors, behavior, experience, and methodology and theory. The trend and patterns of knowledge formation alongside the eight themes are described, visualized, and discussed by two periods of 1978–1999 and 2000-May 2019. The general observation is that despite the remarkable development of family tourism in the past decades, extant research still lacks depth and scope. The seven articles in this collection, by addressing diversified market, various types of attractions, and employing different methods, may enrich the landscape of family tourism.

摘要

本文采用文献计量学的方法，旨在揭示家庭旅游的研究主题在过去四十年中的变化与发展。通过分析，文章总结出家庭旅游研究聚焦在以下八个类别:基础性概念，家庭角色，目的地，家庭结构与功能，心理因素，游客行为，旅游体验以及理论与方法。作者将131篇文章划分为两个阶段 (1978-1999与2000-2019)，并对不同阶段的主题发展进行了可视化分析与探讨。结果表明，尽管近些年来家庭旅游研究数量日益增多，但是研究依然缺乏广度与深度。本期的七篇文章基于不同的研究方法，通过讨论多元化的市场类别，不同类别的目的地和吸引物等来丰富家庭旅游研究的视角。

Introduction

Humans' closest and most significant emotional bonds are formed with their children and families, hence why family lies at the center of social activity (Yeoman, 2008). Contemporary society has placed increasing emphasis on family togetherness at home and during leisure time, including on holidays (Kluin & Lehto, 2012). Family holidays provide an effective means of bonding, during which family members can spend quality time and create collective memories (Carr, 2011; Lehto, Choi, Lin, & MacDermid, 2009). Traveling has become more of a necessity than a luxury for families, spurring the development of a large tourism sector (Lehto et al., 2009). According to Agoda (2018) data on family travel trends, family travel has become popular around the world. For example, in Australia, 71% of travelers reported having traveled with their core family (i.e. parents & children) in 2017; these proportions reached 81% and 86% in Indonesia and Thailand, respectively.

Given the rapidly expanding family travel market, tourism scholars have noticed the ballooning popularity of related tourism in recent years. According to Wu, Wall, Zu, and Ying (2019), family tourism research has thus far focused on three major streams: family travel decision making, the benefits of travel, and family travel experiences. To gain a deeper understanding of the intellectual, social, and conceptual structures of family tourism research in English and Chinese, a bibliometric analysis was conducted as part of this editorial note.

Sample articles were gathered from various databases. English-language materials were collected via EBSCO, Elsevier, Sage, ProQuest, and Taylor & Francis. Chinese-language materials were drawn from China National Knowledge Infrastructure (CNKI); to ensure the quality of articles being reviewed, only those published in Chinese Social Sciences Citation Index (CSSCI) journals were included. Based on keyword searches for 'family tourism,' 'family vacation,' 'family holiday,' 'family leisure,' 'family travel,' 'children-parent tourism,' 'adolescent-parent tourism,' 'father/mother tourism/travel,' 'family life cycle' and 'family function,' 228 articles were returned. We carefully reviewed all articles and discarded several irrelevant studies, such as those pertaining to family business and at-home family leisure. Ultimately, 131 manuscripts published from 1978 to May 2019 were included in our sample. The PDF and RIS versions of each article were downloaded into separate files for further analysis.

We then entered pertinent information from each paper into a Microsoft Excel spreadsheet, namely the article title, journal name, author(s), year of publication, authors' national and university affiliations, keywords, data collection and analysis methods, and major findings. We further scrutinized the data to rectify potential coding issues prior to analysis. Duplicate information such as 'Heike Schänzel' and 'Heike A. Schänzel' (representing the same researcher) and 'AUT University' and 'Auckland University of Technology' (representing the same institution) were modified for consistency in the data cleaning process. Moreover, as our efforts focused on the thematic development of family tourism research taking keywords as significant trend indicators, terms representing identical/similar concepts (e.g. 'family' vs. 'families'; 'FLC' vs. 'family life cycle') were standardized to facilitate analysis.

Our formal analysis involved two steps. First, we conducted descriptive analysis to craft an overall picture of the evolution of family tourism research. Information such as the volume of research output by year and by journal, research collaborations, topics of interest, and chosen research methods were organized. In the second stage, we used VOSviewer to examine the relationships and interactions among research topics and the development of research themes based on co-word analysis.

A profile of publications

The number of articles on family tourism published from 1978 to May 2019 is shown in Figure 1. The earliest publications appeared in 1978, written by Jenkins (1978) and Myers and Moncrief (1978); both articles investigated decision making. Then, no new publications emerged until 1980. As indicated in Figure 1, academia has seen an upsurge of publications on family tourism since 2009.

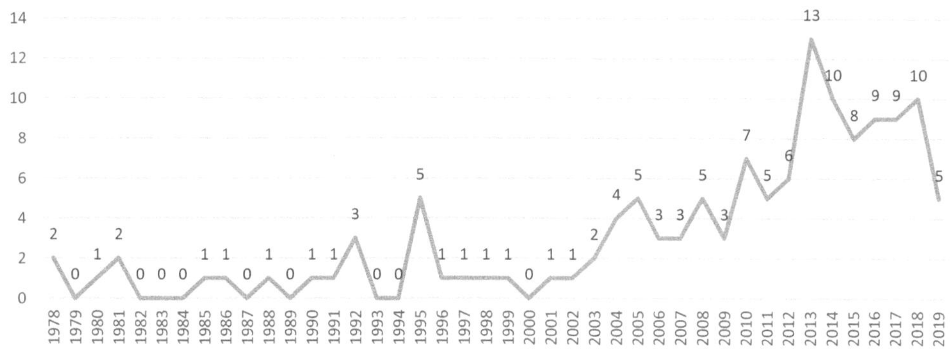

Figure 1. Number of publications by year (1978–May 2019).

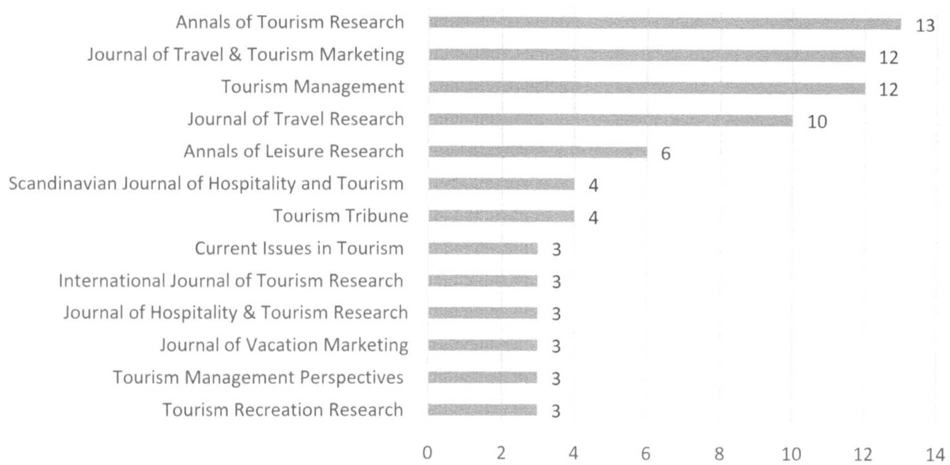

Figure 2. Journal outlets of publications.

The chosen 131 family tourism–related articles (see Appendix) were published across 54 journals; the highest proportion of articles collectively appeared within 13 titles (Figure 2). *Annals of Tourism Research* (ATR) was home to the largest number of publications (13), followed by *Journal of Travel and Tourism Marketing* (JTTM; 12), *Tourism Management* (TM; 12), and *Journal of Travel Research* (JTR; 10). Not surprisingly, most articles appeared in tourism-focused journals. Further investigation revealed that most articles (93) were included in general tourism journals such as JTR and TM as classified by Wu and Wall (2016); only seven were found in business and economics journals such as *Journal of Business Research* (JBR) and *Journal of Marketing Management* (JMM).

Authors' affiliations are listed in Figure 3. Our sample of 131 articles was collectively authored by scholars from 136 institutions around the world, with the largest number of researchers affiliated with Purdue University (26), followed by Zhejiang University (12), Auckland University of Technology (10) and The Hong Kong Polytechnic University (10).

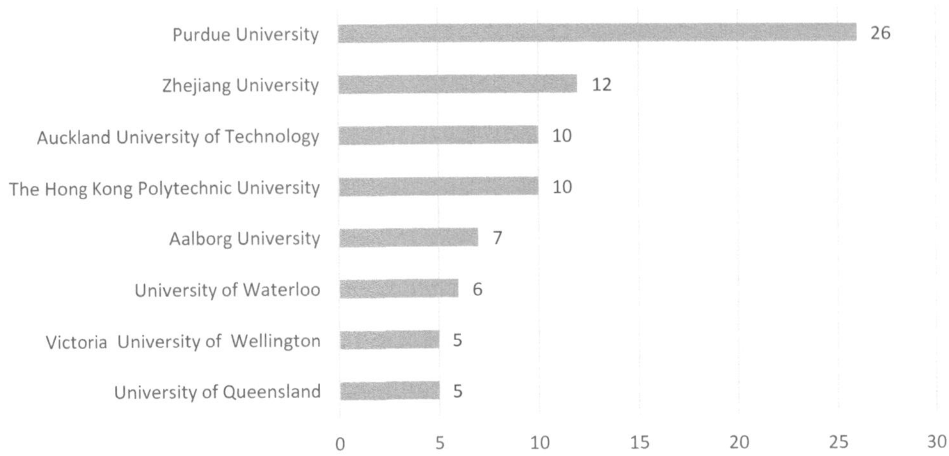

Figure 3. Publications by authors' affiliations.

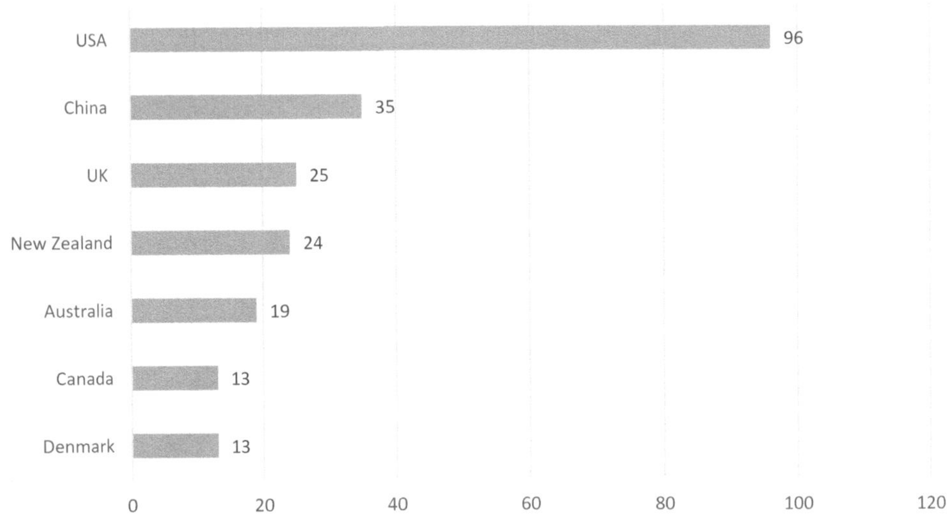

Figure 4. Publications by country of authors' affiliations.

Authors' countries of affiliation are depicted in Figure 4. Only articles written in English (121) were analyzed for this aspect. Most authors were affiliated with universities in the United States (96), followed by the Greater China region (including Hong Kong, Macau, and Taiwan; 35), the United Kingdom (25), New Zealand (24), and Australia (19).

The selected articles were authored by a total of 294 scholars. Figure 5 lists the names of scholars who published at least three articles on family tourism during the timeframe of interest. The landscape of family tourism authorship exhibited clear distinctions by gender and region. Among the seven scholars who published more than three articles, six were women and five were Asian. Among all 294 contributors, 21 published two articles and the others published one, implying that most scholars simply happened to consider this topic; only a handful of researchers appeared to focus on family tourism in particular.

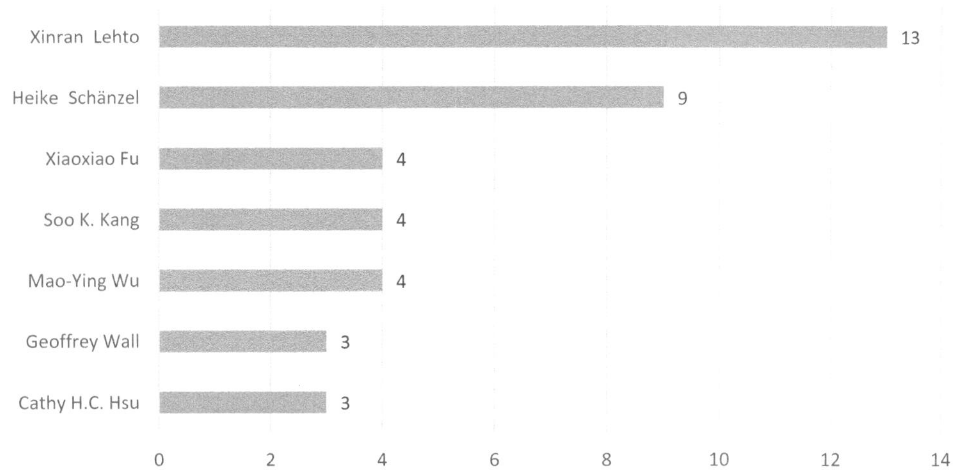

Figure 5. Publications by authors.

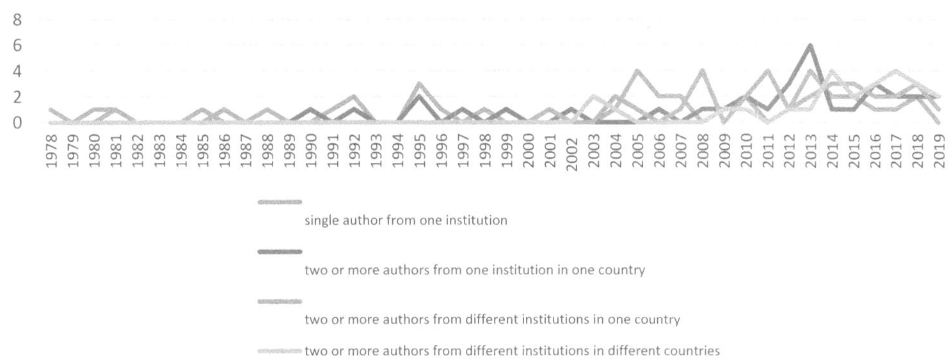

Figure 6. Collaboration patterns by year.

Xinran Lehto from Purdue University and Heike Schänzel from Auckland University of Technology assumed leading roles in knowledge development.

Following Okumus, Koseoglu, and Ma (2018) approach to analyzing collaboration patterns, four types of research collaboration were identified in our sample: a single author; two or more authors from one institution in one country; two or more authors from different institutions in one country; and two or more authors from different institutions in different countries. As illustrated in Figure 6, the first and third collaboration patterns resulted in 36 and 37 publications respectively, 33 papers were featured by the second pattern, and the remaining 25 papers were characterized by the fourth pattern. Therefore, despite evident growth since 2009, international collaboration on family-related tourism research remains insufficient.

Research design

Ruhanen, Weiler, Moyle, and McLennan (2015) suggested classifying theoretical and/or methodological approaches to tourism research into four types: empirical study,

Table 1. Methodological Approaches.

Approach	Details		Count	Percentage
Empirical study	Collection and analysis of qualitative or quantitative data based on direct or indirect observation and experience	Quantitative	62	47.3%
		Qualitative	45	34.4%
		Mixed methods	8	6.1%
Literature review	Critical review of literature on a specific topic		11	8.4%
Historical perspective/ Discussion/ Opinion piece	Analysis and discussion of previous studies to reveal research deficiencies or to make recommendations for future research		4	3.0%
Theory (model) building	Development of a model for a particular research topic		1	0.8%

literature review, theory (model) building, and historical perspective/discussion/opinion piece. Following a similar rule, the methodological approaches used in family tourism studies are summarized in Table 1. Empirical studies (*n* = 115, 87.8%) dominated the sample, followed distantly by literature reviews (*n* = 11, 8.4%). Scholars in this sample (exclude four discussion/opinion piece) performed survey research most often (*n* = 47, 37%) followed by interviews (*n* = 34, 26.8%). Less common designs included the use of secondary data (*n* = 26, 20.5%), mixed methods (*n* = 16, 12.6%), and netnography (*n* = 4, 3.1%).

Compared with the general methodological pattern of tourism research in which a quantitative–deductive approach is most popular, a higher proportion of studies in family tourism relied on qualitative data. Yet most qualitative work, informed by post-positivism, adopted deductive methods to analyze text by counting the frequency of themes while ignoring the nuanced contextual meanings of narratives.

In terms of geography, apart from 14 articles that did not provide such information, the United States (*n* = 28), China (including Hong Kong, Macau, and Taiwan; *n* = 17), the United Kingdom (*n* = 11), and Australia (*n* = 9) were the most common areas of interest. This distribution was largely consistent with that of authors' affiliated countries, suggesting scholars' genuine interest in their local market. Among the 107 articles in English, more than half (*n* = 62) studied families in the same country as the authors' affiliated institutions, and 27 considered families in the same country as one of the authors' affiliated institutions. Eighteen studies were conducted in countries different from the authors' affiliated universities.

Thematic development

Co-word analysis was next conducted to understand the thematic evolution of family tourism research from 1978 to May 2019. After collecting all articles' keywords, we used VOSviewer to develop a holistic view of the interconnections among keywords by visualizing co-word networks (Figure 7). In Figure 7, the size of a bubble represents the frequency of each keyword's appearance, and the line thickness indicates the strength of co-occurrence ties (Leung, Sun, & Bai, 2017). The five keywords receiving the most attention were 'children,' 'decision making,' 'family vacation,' 'family travel,' and 'family.' Further exploration unveiled eight themes capturing the structure of sub-domains in family tourism research throughout the study period: 1) *foundations*, namely the general concept of family tourism–related keywords (e.g. 'family travel' and 'family vacation'); 2)

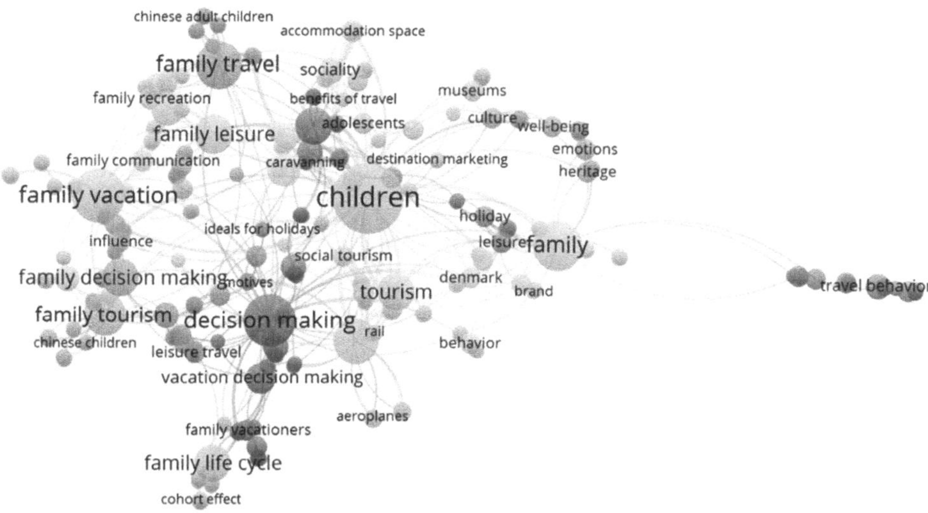

Figure 7. Co-word network: 1978–May 2019.

family roles, or the roles that research participants played in their family (e.g. parents or children); 3) *place*, referring to a specific research area or particular attraction; 4) *family structure and function*, highlighting the family's function in a society and the influence of the family life cycle; 5) *psychological factors*, including tourists' travel motivations, emotions, or satisfaction; 6) *behavior*, particularly what tourists do and how they behave before, during, and after travel; 7) *experience*, such as tourists' various travel experiences; and 8) *methodology and theory*, describing the method(s) or theory(ies) used in a study.

Knowledge creation in family tourism began to rise in 2000 as the publication volume increased and the topic area broadened. From 1978 to 1999, only 22 family tourism–related articles were published in 12 journals. Content analysis of keywords and the eight relevant themes is shown in Table 2; the co-word network is depicted in Figure 8. Different from the overall co-word network, studies conducted in 1978–1999 largely focused on behavioral aspects of family tourism with decision making receiving the most attention. This topic was often studied in the context of family travel/vacations and couples/spouses. The second most frequent keyword was 'family life cycle,' particularly with respect to its influence on tourist expenditure or travel patterns. The cohort effect was also considered part of the family life cycle. 'Children' was the third most common keyword and tended to be explored relative to overall family travel satisfaction, experiences at beach-based attractions, and children's influence on family travel decision making.

Consistent with findings from previous research, behavior was the most thoroughly studied area, including decision making and consumption behavior. The economic value of this market sector clearly captured scholars' interest during the first wave of family tourism research (i.e. 1978–1999), hence their focus on *who* contributes to family vacation decisions (e.g. the influence of each family member as suggested by secondary data). Yet studies on *how* and *why* certain decisions were made remained limited. In addition, most research in this period focused on the pre-trip stage with comparatively little attention to during- and post-trip behavioral responses such as family members' experiences, attitudes, and satisfaction.

Table 2. Keywords: 1978–1999.

Theme	Sub-theme	Keywords (Frequency) (1978–1999)
Behavior (28)	Decision making (14)	decision making (6); family decision making (2); tourist group decision making (1); travel decision (1); consumption decision making (1); decision and influence structure (1); decision making modes (1); lodging choice (1)
	Consumption behavior (9)	consuming behavior preference (2); family clients (1); family consuming struct (1); family tourists consume (1); hotel expenditures (1); income elasticities (1); tourism demand (1); tourist expenditure (1)
	General (5)	tourism/tourists/travel behavior (4); travel pattern (1)
Foundation (24)		family travel (6); family vacation (5); tourism (3); holidays (2); overseas travel (2); disadvantaged families (1); leisure travel (1); types of vacation (1); household vacation travel (1); tours (1); benefits of tourism (1)
Family Roles (15)	Children (9)	children (4); children influence (2); children and tourism (1); teenage child (1); image of tourists (1)
	Parents/Spouses/Couples/Other adults (6)	parent (2); couple (1); gender (1); spouses (1); international travel agent (1)
Place (7)	Countries/Cities (6)	USA (1); Europe (1); Mexico (1); German residents (1); Multi-nation (1); (country) comparison (1)
	Attractions (1)	beaches (1);
Psychological Factors (5)		attitudes (2); cultural identity (1); satisfaction (1); geopolitical identity (1)
Family Structure and Functions (6)		family life cycle (5); cohort effect (1)
Methodology and Theory (6)		children's drawings (1); empirical evidence (1); leisure theory (1); cross sectional (1); longitudinal approach (1); phenomenological perspective (1)
Experience (1)		women holiday experience (1)

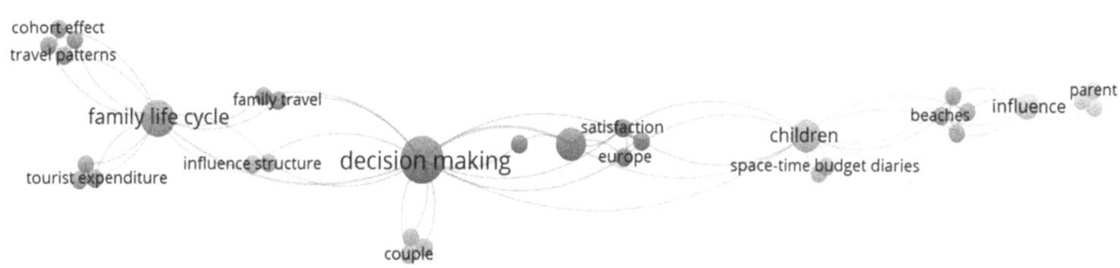

Figure 8. Co-word network: 1978–1999.

Children, despite appearing in more keywords than adults, were seldom studied as active agents but instead passive 'others' during this research period. Given scholars' emphasis on decision making, the degree to which children could influence adults' travel decisions seemed to be the center of attention. According to the prevalence of place-themed keywords, studies from 1978 to 1999 generally adopted a Western orientation specific to North America and Europe. This pattern mostly coincides with Xiao and Smith (2006) notes regarding the geographical focus of tourism research per their index analysis of ATR.

From 2000 to 2019, 109 family tourism–related articles were published in 54 journals. Compared to the first wave of research, the number of relevant keywords in the second wave increased dramatically and the co-word network became more complicated, thus implying richer knowledge production around family tourism. The keyword content analysis is shown in Table 3, and the accompanying co-word network appears in Figure 9.

The co-word network pattern in the second period was similar to the overall network. As shown in Figure 9, the most prominent bubble was 'children,' which was strongly linked to 'family travel,' 'family holiday,' and 'family vacation' along with several small bubbles. Therefore, during this period, the keyword 'children' continued to garner considerable academic attention and extended to various research topics. The influence of this market segment expanded in kind. This trend has been confirmed in other studies (e.g. Blichfeldt, Pedersen, Johansen, & Hansen, 2011) revealing that children have come to exert a growing impact on family travel decision making given the multidimensional nature of family travel (Kozak, 2010).

The keyword 'family leisure,' which was not observed during the first period, emerged as an important focal topic during the second research period. This type of leisure began to be examined in connection with 'family vacation' and 'family functioning,' with the latter being a new keyword in this period as well. For a family to function well, spending time together on leisure activities is critical (Major, Klein, & Ehrhart, 2002). Such quality time can also greatly affect family functioning through cohesion (Lehto et al., 2009); because tourism is a major component of leisure activities (Zabriskie & McCormick, 2001), family leisure assumed a prominent place in the network. Moreover, as indicated in Figure 9, 'decision making' and 'family life cycle' were common research foci in both periods, revealing the market-driven nature of family tourism (Lehto et al., 2009).

The keyword classification for the second research wave appears in Table 3. Only keywords that were cited repeatedly and fell within the eight identified themes were included. The structure of author-provided keywords in the second period reflects several current trends in family tourism.

Table 3. Keywords: 2000–May 2019.

Theme	Sub-theme	Keywords (Frequency) (2000–May 2019)
Foundation (129)	General (61)	family (17); family travel (9); family tourism (8); tourism (4); travel/tourism benefits (3); social tourism (3); (destination/tourism) marketing (3); accommodation (2); destination management (1); Chinese family (1); family characteristics (1); family reunion travel (1); market investigation (1); market segmentation (1); social care (1); social connectedness (1); target market selection (1); tourism mobility (1); tourism products (1); tourism sector (1)
	Family vacation/Holiday (34)	family holiday (14); family vacation (13); vacation (2); vacation planning (1); vacation effect (1); vacation benefits (1); vacation functions (1); vacation marketing (1)
	Types of tourism (18)	adventure tourism (2); culture (2); babymoon tourism (1); business travel (1); coastal mass tourism (1); caravanning (1); dark tourism (1); ecotourism (1); film tourism (1); group package tour (1); medical tourism (1); nature-based adventure tourism (1); niche tourism (1); wedding tourism (1); wildlife tourism (1); festival (1)
	Family leisure/Recreation (16)	family leisure (7); core and balance family leisure (3); family recreation (2); family time (2); Chinese leisure tourists (1); leisure travel (1)
Family Roles (88)	Parents/Spouses/ Couples/Other adults (36)	parents (4); gender (4); fathers (3); dyadic consensus (2); parental style (2); mother (2); couple leisure (1); couple roles (1); dual-earner households (1); dual-parent family (1); conflict and influence tactics (1); conflict level (1)

(*Continued*)

Table 3. (Continued).

Theme	Sub-theme	Keywords (Frequency) (2000–May 2019)
		conflict resolution (1); disagreement-resolution strategy (1); dyadic (dis)agreement (1); family communication (1); family conservation (1); grandparents (1); emerging adulthood (1); women (1); perceived parental power (1); spouses (1); young adults (1); Chinese parents (1); elderly travel (1)
	Children (36)	children (22); children's influence (3); travel with children (1); (disabled children) access (1); disability (1); family with disable children (1); child-psychology (1); children and adult comparisons (1); children geographies (1); children health (1); children's perspectives (1); crying babies (1); Chinese children (1)
	Others (16)	Adolescent(s) (2); family (member) roles (3); visiting friends and relatives (2); adolescent Internet mavens (1); Chinese adolescents (1); household roles (1); single-parent family (1); work-family conflict (1); young tourists (1); young (1); role of family vacation (1); role distribution (1)
Behavior (52)	Decision making (28)	(family/vacation) decision making (27); the role of women in decision making (1);
	General (16)	(tourists/travel/vacation) behavior (6); consumer behavior (4); sociality (3); socialization (1); consumer socialization (1); travel learning (1)
	Consumption behavior (8)	tourism/travel expenditure (3); travel purchase (2); tourism consumption (1); family buying behavior (1); individual choice behavior (1)
Psychological Factors (35)		motives/motivation (10); (emotional/subjective/family) well-being (5); family satisfaction (4); family leisure satisfaction (2); emotion (2); customer satisfaction and loyalty (1); authenticity (1); bonding (1); family dynamics (1); family fear (1); happiness (1);

(*Continued*)

Table 3. (Continued).

Theme	Sub-theme	Keywords (Frequency) (2000–May 2019)
		mood (1); psychological status (1); social identities (1); social relationships (1); travel attitude (1); desires (1)
Place (28)	Country/Cities (19)	Malaysia (3); Denmark (2); Asian (1); Australia (1); Beijing (1); China tourism (1); Germany (1); Greece (1); Mexico (1); New Zealand (1); Romanian consumers (1); Scotland (1); Singapore (1); Turkish culture (1); UK (1); US (1)
	Attractions (9)	heritage (3); museums (2); art museums (1); destination resort (1); family resort (1); Legoland (1)
Methodology and Theory (2)		qualitative research/method (3); whole family approach (2); generalized linear model (1); bibliometric analysis (1); children travel surveys (1); consumption theory (1); core and balance model (1); correlation analysis (1); critical incident technique (1); decision-making models (1); destination research (1); discrete choice model (1); family studies research (1); family systems theory (1); in home survey (1); interpretative phenomenological analysis (1); leisure Motivation Scale (1); literature review (1); multilevel analysis (1); photo-elicitation interviews (1); self-determination theory (1); social power theory (1); reversal theory (1)
Experience (22)		(destination/holiday/leisure/ tourists) experience (12); holiday/vacation activities (5); leisure activities (1); experiential learning (1); overseas educational travel experiences (1); social experience (1); whitewater rafting (1)
Family Structure and Functions (16)		family life circle (5); family functioning (5);

(*Continued*)

Table 3. (Continued).

Theme	Sub-theme	Keywords (Frequency) (2000–May 2019)
		family adaptation and cohesion (1); family bonding (1); family structure (1); group dynamics (1); link function (1); motivations shift map of family life cycle (MSMFLC) (1)

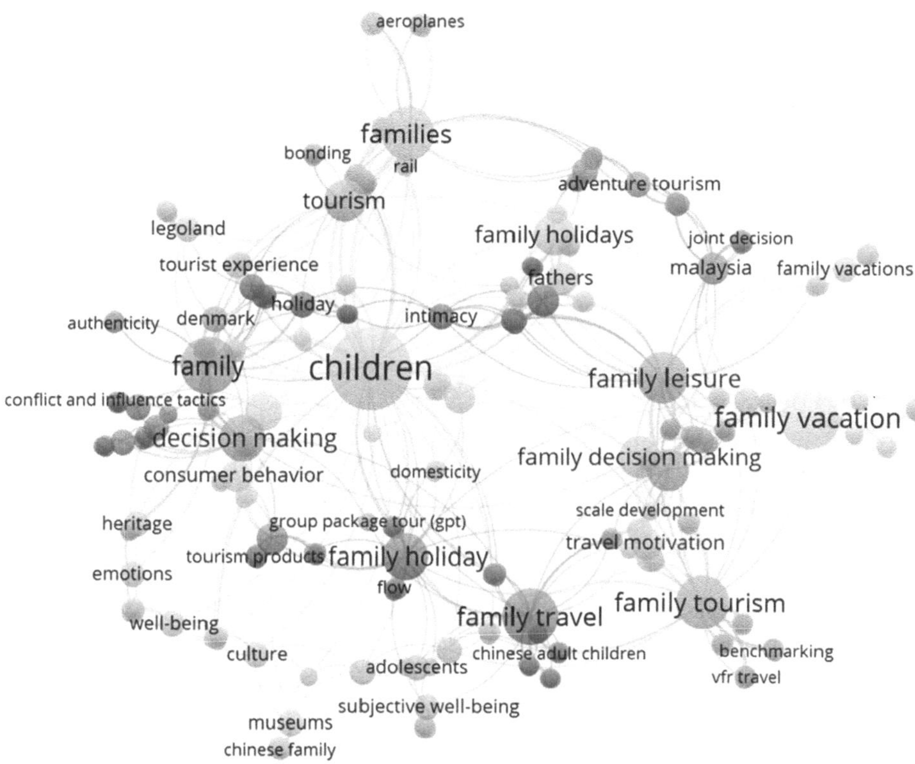

Figure 9. Co-word network: 2000–2019.

First, although 'family vacation/holiday' appeared in the first period, this area expanded to incorporate additional topics such as the benefits and influence of vacations/holidays. Scholars also began to notice the meaning and importance of tourism. Furthermore, as family members might choose different destinations for different purposes, studies on various types of tourism became increasingly diversified (e.g. keywords such as 'adventure tourism'). Kidron (2013) recognized this trend and pointed out that family tourism should not be treated primarily as mass tourism; instead, scholars should consider emerging forms of tourism such as ecotourism. Researchers should also recognize that no particular type of tourism has attracted more attention than others.

Second, 'social tourism' is another new topic that appeared several times during the second sub-period, presumably due to shifts in the social environment. As an emerging form of tourism and an aspect of social policy, social tourism targets low-income families and intends to make tourism accessible to the less fortunate (Minnaert, Maitland, & Miller, 2011). This trend suggests that disadvantaged families are no longer socially excluded from family tourism research.

Third and unexpectedly, the prevalence of psychology-related topics has grown tremendously, specifically those involving motivation and satisfaction. These topics offer myriad marketing implications for family tourism. However, compared with tourists' decision making and behavior, this theme remains relatively narrow and can certainly be developed in the future.

Fourth, as mentioned earlier, children have come to play active roles in family tourism. Recent studies have also begun to consider children with disabilities, as evidenced by keywords such as 'access' and 'disabled children.' These novel focal areas align somewhat with the 'disadvantaged family' notion that emerged during the first research period as well as the 'social tourism' trend highlighted above. Adult-related keywords also grew during the second research wave, particularly on matters of inter-partner negotiation (e.g. 'conflict' or 'disagreement resolution') and the impacts of gender roles in family (e.g. the keywords 'father,' 'mother,' and 'gender' were listed more often). 'Grandparents' and 'elderly travel' also became more common, suggesting senior family members' tourism involvement. These patterns imply that family tourism research is beginning to incorporate more participant groups to examine research questions from diverse perspectives.

Fifth, experience-related themes have diversified from their prior focus on 'women holiday experience,' more recently addressing topics such as family travel experiences at settings ranging from theme parks (e.g. Johns & Gyimothy, 2002) and museums (e.g. Sterry & Beaumont, 2006) to adventurous areas (e.g. Pomfret, 2018). Sixth, different from the 'places' theme in the first research period, place-related keywords exploded during the second wave to include various 'countries/cities' and 'attractions.' Research attention has begun to turn to Asian countries such as Malaysia, China, and Singapore rather than upholding a Western-centric perspective. Attractions have also shifted from solely beaches to other types, especially culture-related attractions such as museums. This growth may be partially attributable to the growing importance of education tourism (Ritchie, Carr, & Cooper, 2003). According to Falk, Ballantyne, Packer, and Benckendorff (2012), tourism has been recognized as a valuable form of experiential learning; as such, parents may begin to favor educational attractions such as museums and heritage sites when traveling with family. Finally, the proliferation of keywords related to theory and analytical techniques suggests that research in the second period involved more diverse methodologies and stronger methodological support for family tourism research.

Remaining issues

The above analysis suggests that despite an upsurge of family tourism research in the past decades, academic endeavor in this area still lacks depth and scope. In addition to the imbalanced contribution in terms of authors' gender and affiliated countries, as well as region of study interest, we also noticed that although 'children' as a keyword has been the center of co-word analysis, they were rarely treated as active agents, as indicated by the small number of research with children as the research participants. Existing research on family tourism is largely confined to conventional tourism destinations such as beach and cultural attractions, while tourists' interface with more diversified and emerging destinations/attractions is missing. Another observation from the thematic analysis is the employment of diversified and novel methodology. While research method is only the means to achieve research objectives, using novel ways of data collection, e.g. real time monitor of tourists' spatial behavior and emotion, may provide richer information to tackle broader scope of research problems.

In response to growing interest in family tourism from the academia and the dramatic development of this market in China, and to address the remaining issues in family tourism research, we compiled this thematic collection of research. The special issue includes seven articles from a total of 18 authors affiliated with nine institutions: four from the United States, three from China, and one each from Australia and Canada. Five of the seven articles were written by women-only research teams, which is relatively uncommon in general tourism scholarship but reflects gendered authorship in work on family tourism.

Among the seven articles, four involved children tourists, indicating the increasingly important role of these family members as active agents in family consumption, including family vacations. Huang and Li (2020) investigated Chinese adolescents' overseas study tours from the perspectives of adolescents and their parents using data gathered through in-depth interviews. Adolescents' major activity types during overseas study tours and their motivations for participating in these tours were identified and interpreted within the context of China's modernization, globalization, single-child policy, and traditional cultural values.

Gao, Havitz, and Potwarka (2020) explored the influence of family vacations on Chinese adolescents' subjective well-being based on survey data. Findings revealed that family vacations positively affected adolescents' subjective well-being, although this impact diminished soon after they returned to school. Using children's drawings of their favorite family holidays along with open-ended interview questions, Khoo-Lattimore and Yang (2020) explored middle-class Malaysian Chinese children's construction of family holidays. Their results highlighted the collectivistic Confucian values of family and education during family holidays. Children's conceptualizations of a 'fun holiday' were also presented. Huang, Zhang, and Ihnatoliova (2020) incorporated methodologies in time geography to explore children tourists' spatial memory bias by matching their GPS-recorded activity paths to their path recall reported on a survey. Two types of memory bias were identified: false memories and memory lost.

The other three articles focused on parents in family travel. Lehto, Fu, Kirillova, and Bi (2020) considered Chinese parents' perceived attractiveness of overseas summer camps from internal and external perspectives using survey data. Several unique camp attribute preferences and motives, such as camp location and educational benefits, were identified. Cai, Wang, and Zhang (2020) attempted to deconstruct the link between vacation travel, marital satisfaction, and subjective well-being based on in-depth interviews. They found vacation travel to enhance marital satisfaction along with individuals' subjective well-being. Four drivers were specifically identified as promoting marital satisfaction, namely mutual devotion, reignited passion, strengthened bonds, and open communication. Drawing upon the growing popularity of P2P accommodations, Lin (2020) investigated families' key considerations when choosing Airbnb rentals during family trips. A unique experience, facilities and space, online reviews, location, and friends' recommendations were highlighted as major concerns.

By presenting this special issue on Family, Children, and Tourism in *Journal of China Tourism Research*, the Guest Editors hope to pique the interest of scholars, regardless of cultural/ethnic, geographical, disciplinary, or methodological backgrounds, to engage in the identified thematic areas of research. We would also like to thank the 18 authors and nearly 20 anonymous reviewers who have contributed to the making of this thematic collection.

Disclosure Statement

No potential conflict of interest was reported by the authors.

References

Agoda. (2018). Asian travelers leading the family travel surge. Agoda releases findings from its global 'Family Travel Trends 2018′ research. Retrieved from https://www.agoda.com/press/family-travel-trends-2018?cid=–218.

Blichfeldt, B. S., Pedersen, B. M., Johansen, A., & Hansen, L. (2011). Tweens on holidays. In-situ decision-making from children's perspective. *Scandinavian Journal of Hospitality and Tourism*, *11*(2), 135–149.

Carr, N. (2011). *Children's and families' holiday experience*. London: Routledge.

Falk, J. H., Ballantyne, R., Packer, J., & Benckendorff, P. (2012). Travel and learning: A neglected tourism research area. *Annals of Tourism Research*, *39*(2), 908–927.

Jenkins, R. L. (1978). Family vacation decision-making. *Journal of Travel Research*, *16*(4), 2–7.

Johns, N., & Gyimothy, S. (2002). Mythologies of a theme park: An icon of modern family life. *Journal of Vacation Marketing*, *8*(4), 320–331.

Kidron, C. A. (2013). Being there together: Dark family tourism and the emotive experience of co-presence in the holocaust past. *Annals of Tourism Research*, *41*, 175–194.

Kluin, J. Y., & Lehto, X. Y. (2012). Measuring family reunion travel motivations. *Annals of Tourism Research*, *39*(2), 820–841.

Kozak, M. (2010). Holiday taking decisions – The role of spouses. *Tourism Management*, *31*(4), 489–494.

Lehto, X. Y., Choi, S., Lin, Y. C., & MacDermid, S. M. (2009). Vacation and family functioning. *Annals of Tourism Research*, *36*(3), 459–479.

Leung, X. Y., Sun, J., & Bai, B. (2017). Bibliometrics of social media research: A co-citation and co-word analysis. *International Journal of Hospitality Management*, *66*, 35–45.

Major, V. S., Klein, K. J., & Ehrhart, M. G. (2002). Work time, work interference with family, and psychological distress. *Journal of Applied Psychology*, *87*(3), 427–436.

Minnaert, L., Maitland, R., & Miller, G. (2011). What is social tourism. *Current Issues in Tourism*, *14*(5), 403–415.

Myers, P. B., & Moncrief, L. W. (1978). Differential leisure travel decision-making between spouses. *Annals of Tourism Research*, *5*(1), 157–165.

Okumus, B., Koseoglu, M. A., & Ma, F. (2018). Food and gastronomy research in tourism and hospitality: A bibliometric analysis. *International Journal of Hospitality Management*, *73*, 64–74.

Ritchie, B. W., Carr, N., & Cooper, C. P. (2003). *Managing educational tourism* (Vol. 10). Bristol: Channel View.

Ruhanen, L., Weiler, B., Moyle, B. D., & McLennan, C. L. J. (2015). Trends and patterns in sustainable tourism research: A 25-year bibliometric analysis. *Journal of Sustainable Tourism*, *23* (4), 517–535.

Sterry, P., & Beaumont, E. (2006). Methods for studying family visitors in art museums: A cross-disciplinary review of current research. *Museum Management and Curatorship, 21*(3), 222–239.
Wu, M. Y., & Wall, G. (2016). Chinese research on family tourism: Review and research implications. *Journal of China Tourism Research, 12*(3–4), 274–290.
Wu, M. Y., Wall, G., Zu, Y., & Ying, T. (2019). Chinese children's family tourism experiences. *Tourism Management Perspectives, 29*, 166–175.
Xiao, H., & Smith, S. L. J. (2006). The making of tourism research: Insights from a social sciences journal. *Annals of Tourism Research, 33*(2), 490–507.
Yeoman, I. (2008). *Tomorrow's tourist: Scenarios and trends*. Oxford: Elsevier.
Zabriskie, R. B., & McCormick, B. P. (2001). The influences of family leisure patterns on perceptions of family functioning. *Family Relations, 50*(3), 281–289.

Appendix: List of Family Tourism Research: 1978-2019

白凯,符国群.(2011).家庭旅游决策研究的中国化理论视角与分析思路.*旅游学刊,26*(12),49–56.
白凯, 张娇, 璩亚杰.(2018).双元孝道对家庭旅游决策行为的影响研究.*旅游科学,32*(1),62–73.
崔庠, 黄安民.(1995).居民家庭旅游消费行为初探.*人文地理,10*(2),37–42.
林裕強, 何嘉琦.(2013).家庭參與生態旅遊之動機與阻礙-不同家庭生命週期之探討.*戶外遊憩研究,26*(4),1–30.
易行健, 盛威, 杨碧云.(2016).家庭收入与人口结构特征对旅游支出的影响效应 – – 基于中国城镇住户调查数据的经验证据.*消费经济,32*(4),38–46.
严艳,周文,张佑印.(2010).基于市场的西安城市家庭旅游决策研究.*地域研究与开发,29*(4),78–81.
殷平, 蔡安雅. (2010).中国家庭出游决策现状及旅游市场营销建议研究 – – 以北京为例.*人文地理,25*(4),137–142.
任明丽, 李群绩, 何建民.(2018).身体状况还是积极心态? – – 关于中国老年家庭出游限制因素的经验分析.*旅游学刊,33*(5),26–43.
谢佳慧,张良.(2018).住房资产, 住房负债与家庭旅游消费.*旅游科学,32*(6),47–64
许春晓, 田媛, 姜漫, & 王洁.(2012).家庭生命周期与旅游态度的关联研究 – – 以长沙市居民为例.*旅游学刊,27*(9),65–72.
Agate, J. R., Agate, S. T., & Birchler, K. (2015). A vacation within a vacation: Children's day programs and parental satisfaction. *Tourism, Culture & Communication, 15*(1), 21–32.
Aslan, N. (2009). An examination of family leisure and family satisfaction among traditional Turkish families. *Journal of Leisure Research, 41*(2), 157–176.
Backer, E., & Lynch, D. (2017). Understanding the proclivity of visiting friends and relatives (VFR) travel across family life cycle stages in Australia. *International Journal of Tourism Research, 19*(4), 447–454.
Backer, E., & Schänzel, H. (2013). Family holidays – Vacation or Obli-cation? *Tourism Recreation Research, 38*(2), 159–173.
Barlés-Arizón, M. J., Fraj-Andrés, E., & Martínez-Salinas, E. (2013). Family vacation decision making: The role of woman. *Journal of Travel & Tourism Marketing, 30*(8), 873–890.
Belch, G. E., Belch, M. A., & Ceresino, G. (1985). Parental and teenage child influences in family decision making. *Journal of Business Research, 13*(2), 163–176.
Bertella, G. (2015). Celebrating the family abroad: The wedding tourism experience. *Annals of Leisure Research, 18*(3), 397–413.
Blichfeldt, B. S., Pedersen, B. M., Johansen, A., & Hansen, L. (2011). Tweens on holidays. In-Situ decision-making from children's perspective. *Scandinavian Journal of Hospitality and Tourism, 11*(2), 135–149.
Bojanic, D. C. (1992). A look at modernized family life cycle and overseas travel. *Journal of Travel & Tourism Marketing, 1*(1), 61–79.
Brey, E. T., & Lehto, X. (2008). Changing family dynamics: A force of change for the family-resort industry? *International Journal of Hospitality Management, 27*(2), 241–248.

Bronner, F., & De Hoog, R. (2008). Agreement and disagreement in family vacation decision-making. *Tourism Management, 29*(5), 967–979.
Buzinde, C. N., & Manuel-Navarrete, D. (2013). The social production of space in tourism enclaves: Mayan children's perceptions of tourism boundaries. *Annals of Tourism Research, 43*, 482–505.
Carr, N. (2006). A comparison of adolescents' and parents' holiday motivations and desires. *Tourism and Hospitality Research, 6*(2), 129–142.
Chen, Y., Lehto, X., & Cai, L. (2013). Vacation and well-being: A study of Chinese tourists. *Annals of Tourism Research, 42*, 284–310.
Choi, H. Y., Lehto, X., & Brey, E. T. (2010). Investigating resort loyalty: Impacts of the family life cycle. *Journal of Hospitality Marketing & Management, 20*(1), 121–141.
Connell, J. (2005). Toddlers, tourism and Tobermory: Destination marketing issues and television-induced tourism. *Tourism Management, 26*(5), 763–776.
Cosenza, R. M., & Davis, D. L. (1981). Family vacation decision making over the family life cycle: A decision and influence structure analysis. *Journal of Travel Research, 20*(2), 17–23.
Cullingford, C. (1995). Children's attitudes to holidays overseas. *Tourism Management, 16*(2), 121–127.
Davidson, P. (1996). The holiday and work experiences of women with young children. *Leisure Studies, 15*(2), 89–103.
Davies, A., & Christie, N. (2018). The experiences of parents with children with disabilities traveling on planes: An exploratory study. *Journal of Transport & Health, 11*, 122–129.
Davies, B., & Mangan, J. (1992). Family expenditure on hotels and holidays. *Annals of Tourism Research, 19*(4), 691–699.
Drenten, J. (2018). When kids are the last to know: Embodied tensions in surprising children with family vacations. *Young Consumers, 19*(2), 199–217.
Durko, A. M., & Petrick, J. F. (2013). Family and relationship benefits of travel experiences: A literature review. *Journal of Travel Research, 52*(6), 720–730.
Feng, X., & Li, M. (2016). Children tourism: A literature review. *Tourism Tribune 31*(9), 61–71.
Fodness, D. (1992). The impact of family life cycle on the vacation decision-making process. *Journal of Travel Research, 31*(2), 8–13.
Fountain, J., Schänzel, H., Stewart, E., & Körner, N. (2015). Family experiences of visitor attractions in New Zealand: Differing opportunities for 'family time' and 'own time'. *Annals of Leisure Research, 18*(3), 342–358.
Frost, W., & Laing, J. H. (2017). Children, families and heritage. *Journal of Heritage Tourism, 12* (1), 1–6.
Fu, X., Huang, Z., Li, Q., & Kirillova, K. (2018). Dissecting Chinese adolescents' overseas educational travel experiences: Movements, representations and practices. *Current Issues in Tourism, 21*(10), 1115–1136.
Fu, X., & Lehto, X. (2018). Vacation co-creation: The case of Chinese family travelers. *International Journal of Contemporary Hospitality Management, 30*(2), 980–1000.
Fu, X., Lehto, X., & Park, O. (2014). What does vacation do to our family? Contrasting the perspectives of parents and children. *Journal of Travel & Tourism Marketing, 31*(4), 461–475.
Gabor, M. R., & Oltean, F. D. (2019). Babymoon tourism between emotional well-being service for medical tourism and niche tourism. Development and awareness on Romanian educated women. *Tourism Management, 70*, 170–175.
Gamradt, J. (1995). Jamaican children's representations of tourism. *Annals of Tourism Research, 22* (4), 735–762.
Gram, M. (2005). Family holidays. A qualitative analysis of family holiday experiences. *Scandinavian Journal of Hospitality and Tourism, 5*(1), 2–22.
Gram, M. (2007). Children as co-decision makers in the family? The case of family holidays. *Young Consumers, 8*(1), 19–28.
Gustafson, P. (2006). Work-related travel, gender and family obligations. *Work, Employment and Society, 20*(3), 513–530.
Hagemann, R. P. (1981). The determinants of household vacation travel: Some empirical evidence. *Applied Economics, 13*(2), 225–234.

Hall, S. M., & Holdsworth, C. (2016). Family practices, holiday and the everyday. *Mobilities, 11*(2), 284–302.
Hay, B. (2017). Missing voices: Australian children's insights and perceptions of family holidays. *Hospitality & Society, 7*(2), 133–155.
Hazel, N. (2005). Holidays for children and families in need: An exploration of the research and policy context for social tourism in the UK. *Children & Society, 19*(3), 225–236.
Hilbrecht, M., Shaw, S. M., Delamere, F. M., & Havitz, M. E. (2008). Experiences, perspectives, and meanings of family vacations for children. *Leisure/Loisir, 32*(2), 541–571.
Hong, G., Fan, J. X., Palmer, L., & Bhargava, V. (2005). Leisure travel expenditure patterns by family life cycle stages. *Journal of Travel & Tourism Marketing, 18*(2), 15–30.
Hornberger, L. B., Zabriskie, R. B., & Freeman, P. (2010). Contributions of family leisure to family functioning among single-parent families. *Leisure Sciences, 32*(2), 143–161.
Hughes, K., Packer, J., & Ballantyne, R. (2011). Using post-visit action resources to support family conservation learning following a wildlife tourism experience. *Environmental Education Research, 17*(3), 307–328.
Hung, W. T., Shang, J. K., & Wang, F. C. (2013). A multilevel analysis on the determinants of household tourism expenditure. *Current Issues in Tourism, 16*(6), 612–617.
Hunter-Jones, P. (2014). Changing family structures and childhood socialization: A study of leisure consumption. *Journal of marketing management., 30*(15–16), 1533–1553.
Jamal, S. A., Aminudin, N., & Kausar, D. R. (2019). Family adventure tourism motives and decision-making: A case of whitewater rafting. *Journal of Outdoor Recreation and Tourism, 25*, 10–15.
Jenkins, R. L. (1978). Family vacation decision-making. *Journal of Travel Research, 16*(4), 2–7.
Ji, J., Anderson, D., Wu, X., & Kang, C. (2014). Chinese family groups' museum visit motivations: A comparative study of Beijing and Vancouver. *Curator: The Museum Journal, 57*(1), 81–96.
Johns, N., & Gyimóthy, S. (2002). Mythologies of a theme park: An icon of modern family life. *Journal of Vacation Marketing, 8*(4), 320–331.
Johns, N., & Gyimóthy, S. (2003). Postmodern family tourism at Legoland. *Scandinavian Journal of Hospitality and Tourism, 3*(1), 3–23.
Kang, S. K., & Hsu, C. H. C. (2004). Spousal conflict level and resolution in family vacation destination selection. *Journal of Hospitality & Tourism Research, 28*(4), 408–424.
Kang, S. K., & Hsu, C. H. C. (2005). Dyadic consensus on family vacation destination selection. *Tourism Management, 26*(4), 571–582.
Kang, S. K., Hsu, C. H. C., & Wolfe, K. (2003). Family traveler segmentation by vacation decision-making patterns. *Journal of Hospitality & Tourism Research, 27*(4), 448–469.
Kennedy-Eden, H., & Gretzel, U. (2016). Modern vacations – Modern families: New meanings and structures of family vacations. *Annals of Leisure Research, 19*(4), 461–478.
Khoo-Lattimore, C., Prayag, G., & Cheah, B. L. (2015). Kids on board: Exploring the choice process and vacation needs of Asian parents with young children in resort hotels. *Journal of Hospitality Marketing & Management, 24*(5), 511–531.
Kidron, C. A. (2013). Being there together: Dark family tourism and the emotive experience of co-presence in the holocaust past. *Annals of Tourism Research, 41*, 175–194.
Kim, S., & Lehto, X. (2013). Travel by families with children possessing disabilities: Motives and activities. *Tourism Management, 37*, 13–24.
Kim, S. S., Choi, S., Agrusa, J., Wang, K. C., & Kim, Y. (2010). The role of family decision makers in festival tourism. *International Journal of Hospitality Management, 29*(2), 308–318.
Kluin, J. Y., & Lehto, X. (2012). Measuring family reunion travel motivations. *Annals of Tourism Research, 39*(2), 820–841.
Koc, E. (2004). The role of family members in the family holiday purchase decision-making process. *International Journal of Hospitality & Tourism Administration, 5*(2), 85–102.
Kozak, M. (2010). Holiday taking decisions – The role of spouses. *Tourism Management, 31*(4), 489–494.

Kozak, M. (2016). Family-based travel narratives: Confirmatory personal introspection of children's interpretations of their journey to three destinations. *Journal of Hospitality and Tourism Management, 29*, 119–125.

Larsen, J. R. K. (2013). Family flow: The pleasures of 'being together' in a holiday home. *Scandinavian Journal of Hospitality and Tourism, 13*(3), 153–174.

Lawson, R. (1991). Patterns of tourist expenditure and types of vacation across the family life cycle. *Journal of Travel Research, 29*(4), 12–18.

Lee, B., Graefe, A., & Burns, R. (2008). Family recreation: A study of visitors who travel with children. *World Leisure Journal, 50*(4), 259–267.

Lehto, X., Choi, S., Lin, Y. C., & MacDermid, S. M. (2009). Vacation and family functioning. *Annals of Tourism Research, 36*(3), 459–479.

Lehto, X., Fu, X., Li, H., & Zhou, L. (2017). Vacation benefits and activities: Understanding Chinese family travelers. *Journal of Hospitality & Tourism Research, 41*(3), 301–328.

Lehto, X., Lin, Y. C., Chen, Y., & Choi, S. (2012). Family vacation activities and family cohesion. *Journal of Travel & Tourism Marketing, 29*(8), 835–850.

Li, M., Wang, D., Xu, W., & Mao, Z. (2017). Motivation for family vacations with young children: Anecdotes from the Internet. *Journal of Travel & Tourism Marketing, 34*(8), 1047–1057.

Liang, Y. W. (2013). Children's influence on purchasing tourism products via the Internet: Parental power versus children's power – The social power perspective. *Journal of Travel & Tourism Marketing, 30*(7), 639–661.

Lin, V. S., Mao, R., & Song, H. (2015). Tourism expenditure patterns in China. *Annals of Tourism Research, 54*, 100–117.

Litvin, S. W., Xu, G., & Kang, S. K. (2004). Spousal vacation-buying decision making revisited across time and place. *Journal of Travel Research, 43*(2), 193–198.

Matthews, B., & Price, L. L. (2013). Travel time as quality time: Parental attitudes to long distance travel with young children. *Journal of Transport Geography, 32*, 49–55.

Michie, D. A. (1986). Family travel behavior and its implications for tourism management: An international study. *Tourism Management, 7*(1), 8–20.

Michie, D. A., & Sullivan, G. L. (1990). The role(s) of the international travel agent in the travel decision process of client families. *Journal of Travel Research, 29*(2), 30–38.

Mikkelsen, M. V., & Stilling Blichfeldt, B. (2015). 'We have not seen the kids for hours': The case of family holidays and free-range children. *Annals of Leisure Research, 18*(2), 252–271.

Mottiar, Z., & Quinn, D. (2012). Is a self-catering holiday with the family really a holiday for mothers? Examining the balance of household responsibilities while on holiday from a female perspective. *Hospitality & Society, 2*(2), 197–214.

Mura, P., & Khoo-Lattimore, C. (2012). Young tourists, gender and fear on holiday. *Current Issues in Tourism, 15*(8), 707–724.

Myers, P. B., & Moncrief, L. W. (1978). Differential leisure travel decision making between spouses. *Annals of Tourism Research, 5*(1), 157–165.

Nanda, D., Hu, C., & Bai, B. (2007). Exploring family roles in purchasing decisions during vacation planning. *Journal of Travel & Tourism Marketing, 20*(3–4), 107–125.

Ndubisi, N. O. (2007). Impact of joint product usage and family structure on joint decision to purchase a vacation by Malaysian spouses. *Journal of Vacation Marketing, 13*(2), 135–147.

Nichols, C. M., & Snepenger, D. J. (1988). Family decision making and tourism behavior and attitudes. *Journal of Travel Research, 26*(4), 2–6.

Nickerson, N. P., & Jurowski, C. (2001). The influence of children on vacation travel patterns. *Journal of Vacation Marketing, 7*(1), 19–30.

Obrador, P. (2012). The place of the family in tourism research: Domesticity and thick sociality by the pool. *Annals of Tourism Research, 39*(1), 401–420.

Oppermann, D. M. (1995). Family life cycle and cohort effects: A study of travel patterns of German residents. *Journal of Travel & Tourism Marketing, 4*(1), 23–44.

Park, O. J., Park, J. K., & Lehto, X. (2008). Service failures and complaints in the family travel market: A justice dimension approach. *Journal of Services Marketing, 22*(7), 520–532.

Pomfret, G. (2018). Conceptualizing family adventure tourist motives, experiences and benefits. *Journal of Outdoor Recreation and Tourism, 28.*
Poria, Y., & Timothy, D. J. (2014). Where are the children in tourism research? *Annals of Tourism Research, 47*, 93–95.
Ram, Y., Uriely, N., & Malach-Pines, A. (2014). Releasing control: Parents on vacation. *International Journal of Tourism Research, 16*(3), 232–240.
Ramer, S. I., Zorotovich, J., Roberson, P. N. E., Flanigan, N., & Gao, J. (2019). Effects of preexisting family dynamics on emerging adult college students' emotions over the course of fall break. *Journal of Destination Marketing & Management.*
Ritchie, J. R. B., & Filiatrault, P. (1980). Family vacation decision-making – A replication and extension. *Journal of Travel Research, 18*(4), 3–14.
Rojas-de Gracia, M. M., & Alarcón-Urbistondo, P. (2016). Toward a gender understanding of the influence of the couple on family vacation decisions. *Tourism Management Perspectives, 20*, 290–298.
Schänzel, H. (2013). The importance of 'social' in family tourism. *Asia-Pacific Journal of Innovation in Hospitality and Tourism, 2*(1), 1–15.
Schänzel, H., & Lynch, P. A. (2016). Family perspectives on social hospitality dimensions while on holiday. *Tourist Studies, 16*(2), 133–150.
Schänzel, H., & Smith, K. A. (2011). The absence of fatherhood: achieving true gender scholarship in family tourism research. *Annals of Leisure Research, 14*(2–3), 143–154.
Schänzel, H., & Smith, K. A. (2014). The socialization of families away from home: Group dynamics and family functioning on holiday. *Leisure Sciences, 36*(2), 126–143.
Schänzel, H., & Yeoman, I. (2014). The future of family tourism. *Tourism Recreation Research, 39* (3), 343–360.
Schänzel, H., & Yeoman, I. (2015). Trends in family tourism. *Journal of Tourism Futures, 1*(2), 141–147.
Shavanddasht, M., & Schänzel, H. (2018). Measuring adolescents' tourism satisfaction: The role of mood and perceived parental style. *Tourism and Hospitality Research, 19*(3), 308–320.
Small, J., & Harris, C. (2014). Crying babies on planes: Aeromobility and parenting. *Annals of Tourism Research, 48*, 27–41.
Smith, V., & Hughes, H. (1999). Disadvantaged families and the meaning of the holiday. *International Journal of Tourism Research, 1*(2), 123–133.
Sterry, P., & Beaumont, E. (2006). Methods for studying family visitors in art museums: A cross-disciplinary review of current research. *Museum Management and Curatorship, 21*(3), 222–239.
Stone, M. J., & Petrick, J. F. (2017). Exploring learning outcomes of domestic travel experiences through mothers' voices. *Tourism Review International, 21*(1), 17–30.
Tagg, D. S., & Seaton, A. V. (1995). The family vacation in Europe. *Journal of Travel & Tourism Marketing, 4*(1), 1–21.
Therkelsen, A. (2010). Deciding on family holidays – Role distribution and strategies in use. *Journal of Travel & Tourism Marketing, 27*(8), 765–779.
Thornton, P. R., Shaw, G., & Williams, A. M. (1997). Tourist group holiday decision-making and behavior: The influence of children. *Tourism Management, 18*(5), 287–297.
Tirasattayapitak, S., Chaiyasain, C., & Beeton, R. J. S. (2015). The impacts of nature-based adventure tourism on children in a Thai village. *Tourism Management Perspectives, 15*, 122–127.
Townsend, J. A., Van Puymbroeck, M., & Zabriskie, R. B. (2017). The core and balance model of family leisure functioning: A systematic review. *Leisure Sciences, 39*(5), 436–456.
Wang, K., Hsieh, A., Yeh, Y., & Tsai, C. (2004). Who is the decision-maker: The parents or the child in group package tours? *Tourism Management, 25*(2), 183–194.
Wang, W., Yi, L., Wu, M. Y., Pearce, P. L., & Huang, S. (2018). Examining Chinese adult children's motivations for traveling with their parents. *Tourism Management, 69*, 422–433.
Watne, T. A., Brennan, L., & Winchester, T. M. (2014). Consumer socialization agency: Implications for family decision-making about holidays. *Journal of Travel & Tourism Marketing, 31*(6), 681–696.

Whittington, A. (2014). Family Vacation 2050: Socially and technologically – driven scenarios of the future of family travel, recreation and tourism. *Tourism Recreation Research, 39*(3), 379–396.

Wu, M. Y., & Wall, G. (2016). Chinese research on family tourism: Review and research Implications. *Journal of China Tourism Research, 12*(3–4), 274–290.

Wu, M. Y., & Wall, G. (2017). Visiting heritage museums with children: Chinese parents' motivations. *Journal of Heritage Tourism, 12*(1), 36–51.

Wu, M. Y., Wall, G., Zu, Y., & Ying, T. (2019). Chinese children's family tourism experiences. *Tourism Management Perspectives, 29*, 166–175.

Yang, F. X., & Lau, V. (2019). Experiential learning for children at World Heritage Sites: The joint moderating effect of brand awareness and generation of Chinese family travelers. *Tourism Management, 72*, 1–11.

Yoo, H., McIntosh, A., & Cockburn-Wootten, C. (2016). Time for me and time for us: conference travel as alternative family leisure. *Annals of Leisure Research, 19*(4), 444–460.

Yu, X., Anaya, G. J., Miao, L., Lehto, X., & Wong, I. A. (2017). The impact of smartphones on the family vacation experience. *Journal of Travel Research, 57*(5), 579–596.

Yun, J., & Lehto, X. (2009). Motives and patterns of family reunion travel. *Journal of Quality Assurance in Hospitality & Tourism, 10*(4), 279–300.

Zalatan, A. (1998). Wives involvement in tourism decision processes. *Annals of Tourism Research, 25*(4), 890–903.

Zhao, X., Guan, H., & Gong, L. (2011). Modeling correlation of family holiday activities based on structural equation model. *Journal of Transportation Systems Engineering and Information Technology, 11*(4), 85–90.

The Grand Tour in the Twenty-first Century: Perspectives of Chinese Adolescents and Their Parents

Zhuowei (Joy) Huang and Qian Li

ABSTRACT
This study investigates Chinese adolescents' overseas study tours from the perspectives of both Chinese adolescents and their parents. Fifty in-depth interviews were conducted and analyzed with 30 adolescents and 20 parents. Data analysis results suggest three primary types of activities in an overseas study tour: sightseeing tours, schooling programs and socializing with the locals. A variety of reasons for overseas study tours from the perspectives of Chinese adolescents (a taste of college education in foreign countries, global perspectives, socialization, independence, novelty and fun) and their parents (child's future plans, child's social development, the lost college dream) are identified. The key findings of this study pinpoint a number of important social, economic, cultural forces in today's China society that shape and are reflected in Chinese adolescents' and their parents' perspectives on overseas study tours, including globalization, China's modernization, China's contemporary single-child family structure and Chinese traditional cultural values.

二十一世纪之'游学大旅行'：中国青少年和父母的游学观点调查

摘要
本研究调查了中国青少年和家长对出国游学旅行的观点。作者对五十名参与者进行了深度访谈，包括30名学生和20名家长。数据分析揭示了游学旅行的三项主要活动：观光旅行、学校教育项目以及与当地人的社交活动。中国青少年参与出国游学旅行包括多方面的原因：体验国外大学教育、建立全球化视野、社交、培养独立性以及新奇与乐趣。中国家长支持孩子出国游学的原因包括：未来教育计划、培养社会能力、寻找失落的大学梦。本研究通过分析中国学生和家长对出国游学旅行的观点映射出当代中国社会的特点，尤其是全球化、中国社会现代化、中国独生子女家庭结构以及中国传统文化之间的碰撞和融合。

Introduction

Traveling abroad for short-term educational and leisure programs (hereafter 'overseas study tours') has become popular in recent years among Chinese adolescents (14–18 years old) from middle-class families. Similar to the Grand Tour in Europe, these adolescents travel to developed countries (e.g. the United States, United Kingdom, etc.) for educational and leisure experiences. It has been reported that over 300,000 Chinese students attended overseas study tours in 2013 (Nan, 2014). Generally speaking, such a tour takes about 2–4 weeks and costs RMB 30,000 to RMB 50,000 (up to $5,000 to $8,300). The majority of them are operationalized by travel agencies and high schools. They travel in groups without their parents' company. Owing to the increasing number of Chinese college students seeking overseas study opportunities in recent years, overseas study tours, perceived as 'mini overseas study', have grown as fast as 30% per year regardless of the constantly increasing price of the tours (Qu, 2013). In the summer of 2015, over 500,000 Chinese adolescents attended such overseas study tours (Zheng, 2015).

While the parents of Chinese adolescents do not physically attend the overseas study tour, they play an important role in the whole decision-making process. Reasons and perspectives regarding the tour, similar or different, can be found from the two generations. Some similarities are expected due to the significant influences of parents on adolescents' opinions (Carlson & Grossbart, 1988; Moschis, 1985). Such influences are even more significant in Chinese families ruled by Confucian ideology that emphasizes the parents' authority and the child's obedience (Chao, 1996). Today's Chinese parents of adolescents are more concerned about their children and tend to exert more influences than previous generations. This is because China's single-child policy since 1979 to 2016 resulted in large numbers of families with only one child (Chan & McNeal, 2003).

On the other hand, the globalization and China's opening-up policy since 1979 have brought dramatic changes to this country on economic, social and cultural aspects. In the last three decades, China has gradually transformed itself from a mandatory planning economy to a modern market economy (Logan, 2011). The fast economic growth and China's increasing integration into the world introduced new and different beliefs, values, cultures and perspectives to Chinese people's lives and their ways of thinking and doing things. Comparing with the time before 1979 when the country suffered the Cultural Revolution and Chinese people had extremely limited mobilities and accessibility to educational resources, today's Chinese have tons of opportunities to travel abroad, interact with diverse cultures, and to have access to educational resources.

Under this backdrop, tourism has been booming in China, domestically and internationally. Despite of an economic activity, the social leveraging power of tourism has emerged in China. As a constitutive proportion of many people's lives, the immersive integration of tourism into China's society has made it reflect the economic, political, social, and cultural development process (Hannam, Butler, & Paris, 2013). It is 'not only an outcome of change, but a catalyst for further change' in China (Ryan & Huang, 2013, p. xix). Therefore, the 'what, how, why' questions about Chinese people travel should not be limited within the tourism setting, but also be investigated in the broad social context. Yet, the fact that tourism can reflect the economic, social and cultural forces in society has been largely overlooked in tourism research. The current study is conducted to fill up this research gap by using Chinese adolescents' overseas study tours to elaborate the interactions between tourism and economic, social and cultural forces in Chinese society. In particular, four questions will be answered in this study.

(1) What are the primary activities of Chinese adolescents' overseas study tours?

(2) Why do Chinese adolescents participate in overseas study tours?
(3) Why do parents support their children (Chinese adolescents) in overseas study tours?
(4) How do economic, social and cultural forces in Chinese society intertwine and influence Chinese adolescents' and their parents' perspectives on and behaviors in an overseas study tour?

Considering that parents exert considerable influences on Chinese adolescents with regard to their opinions, attitudes and behaviors, this study aims to answer the above-mentioned question, and include the perspectives of both parents and children.

Literature review

Overseas travel for education

Globalization has profoundly influenced education all over the world. In the past several decades, a growing number of colleges have developed overseas educational programs to equip students with the knowledge and skills to meet the needs of national and international markets (Pickert, 1992). Overseas educational programs offer students the opportunity to integrate educational experiences into real-world exposure by crossing geographical borders, which has been reflected in tourism research since the early 1990s (e.g. Bosselman, Fernsten, Manning, & Kisseleff, 1989; Dukes, Lockwood, Oliver, Pezalla, & Wilker, 1994; Pizam, Jafari, & Milman, 1991; Thibadoux & Helms, 1989). The review of literature indicates two primary forms of overseas educational programs: study abroad and intern abroad.

Study-abroad programs have proliferated, especially among colleges in the United States and other developed countries since the 1990s (Lee, 2012). These programs vary by duration, activities and organizers. The duration of study abroad programs ranges from short-term field trips (1 to 8 weeks) to semester-long exchange programs (15 to 30 weeks) (Behnke, Seo, & Miller, 2014). Short-term field trips are usually developed by the home college as a combination of a domestic course and a study tour in a foreign country. The study tour is often led by a faculty member who accompanies the students to the destination country and guides the study tour. Students are supervised to conduct academic assignments via the study tour, and afterwards can receive academic credits toward their major (Houser, Brannstrom, Quiring, & Lemmons, 2011). Students have to pay a program fee to attend the short-term field trips. The long-term study-abroad programs are primarily exchange programs, which are usually co-organized by the home college and a host college in another country. It is an unescorted form of study abroad (i.e. no faculty accompanying students). Students normally spend a semester or a whole academic year at a foreign college for academic study. Participating in classes at the host college will be approved for credits by the home college. Students need to pay the tuition fee of the home college to attend exchange programs.

In addition to study-abroad programs, intern-abroad is another important form of overseas educational travel. It bridges the gap between classroom learning and real-world application (Toncar & Cudmore, 2000). Unlike study-abroad programs that mostly take place in college settings, intern-abroad programs are generally managed and hosted by companies, governments, or non-governmental organizations (NGOs).

In general, these organizations provide college students with overseas internships that are aligned with their academic or professional goals (van't Klooster, van Wijk, Go, & van Rekom, 2008). Colleges may partner with those non-academic organizations to help students enroll in intern-abroad programs. The length of a program can vary depending upon the need of the host organization and the interests of students, ranging from a few weeks to a year. Students are placed into work situations either for getting paid or as a volunteer (Wearing, 2001). The internship experience can be approved for credits if the students enroll in an accredited program that is associated with their home college, otherwise not.

Why overseas education?

As noted, overseas educational programs (study abroad and intern abroad) have increasingly gained popularity and interest among college students in the last few decades. Although study abroad or intern abroad is usually not required as a condition for college graduation, the number of students enrolled in such programs is steadily increasing (Redden, 2014). A review of the literature has indicated several factors that boost the overseas educational travel industry. From the standpoint of program providers, colleges tend to offer more study-abroad programs that enable students to face the challenges of globalization. From the perspective of many non-academic organizations (e.g. companies, governments and NGOs), they can globally recruit talented students via intern-abroad programs (Toncar & Cudmore, 2000).

Study-abroad and intern-abroad programs provide college students with a variety of benefits. The review of the literature suggests two salient reasons that were frequently mentioned by college students for study abroad: the improvement of foreign language skills (e.g. Doyle et al., 2010) and the acquisition of cross-cultural knowledge/experience (e.g. Chieffo & Griffiths, 2009; Van Hoof, 2000). Studying abroad is believed to be more beneficial than domestic study in terms of language learning as it offers immersive learning opportunities that are unavailable at home (Doyle et al., 2010). In addition, living in a new culture and interacting with local people can help students gain a better understanding of different culture (Nyaupane, Paris, & Teye, 2011). Research also found that study-abroad programs offer students touristic experiences in which they satisfy their need for leisure travel (Juvan & Lesjak, 2011), although leisure is not typically identified as the primary reason. In summary, learning is regarded as the main purpose for studying abroad from the student's perspective (e.g. Nyaupane et al., 2011; Ritchie, 2003). Similar to reasons to participate in study-abroad programs, mastering a foreign language and increasing cross-cultural knowledge are also indicated by students as important reasons to intern abroad. In addition, interning abroad is perceived as a useful vehicle by college students to increase their career competiveness and opportunities (van't Klooster et al., 2008). The intern-abroad experience can help a participant's résumé stand out more easily among peers and get noticed by recruiters (Toncar & Cudmore, 2000).

The literature review indicates that the majority of existing research on overseas educational travel has primarily focused on college students traveling between developed countries (e.g. Anderson, Lawton, Rexeisen, & Hubbard, 2006; Nyaupane et al., 2011). Recent studies found overseas volunteer tourism programs becoming popular among college students in the last decade (e.g. Lyons, Hanley, Wearing, & Neil, 2012; Sin, 2009). The majority of such tours are from developed countries to developing countries. College students receive educational and other benefits from these tours. The movement of a college student's educational travel across countries is changing from being among developed countries to traveling from developed countries to underdeveloped countries. According to ICEF, over half of the patrons attending educational travel programs nowadays are from developing countries in Asia, such as China, Thailand, India, etc. (ICEF Monitor Report, 2014) and the destinations are developed countries such as the UK, US, Australia, Japan, etc. Tourism research has just started to investigate this growing trend in emerging markets (e.g. Fu, Huang, Li, & Kirillova, 2017), while more investigations into the internal and external forces that influence these travels are needed.

Method

Research design and data collection

This study employed the qualitative research method of in-depth interviews to investigate the driving forces for Chinese adolescents' overseas study tours from the perspectives of children and parents. The qualitative method was chosen primarily because little is known about the above-mentioned question raised in this study (Corbin & Strauss, 2008). Using the qualitative method of in-depth interviews allows this research to answer this question through obtaining first-hand experiences and opinions from informants (Yan & Gray, 1994).

Chinese adolescents who had an overseas study tour in the past three years and their parents were recruited to participate in this study. A recruitment letter was first posted to the parents of qualified adolescents who were on several Chinese online forums with educational travel abroad topics, as well as on popular Chinese social networking platforms (e.g. Qzone, WeChat, Weibo and Renren). The purpose of the research, the interview procedure, as well as the researcher's contact information were stated in the recruitment letter. Once a contact message from a potential informant was received, one researcher followed it up via phone calls and emails. This initial conversation was used to identify the qualification of the potential student informants and to explain the research purpose and procedures. In particular, with recognition of the ethical issues associated with minors, the questions that minor informants were to be asked were provided to the parents and consents were obtained from the parents. With the consent of parents, the students were connected to researchers via parents. Again, the conversation started by fully explaining the research and interview procedures and then asking for consent from minor students. The interviews could only be conducted with the consent obtained from the students and their parents. Interviewers also ensured that each participant understood his or her right during the interview, such as withdraw any time upon their willingness, freely answer the questions, etc. The conversations during the interview used simple language, avoiding academic terms, and participants were encouraged to ask for clarification at any time.

To identify and verify the categories, the strategy of theoretical sampling was used in this study (Coyne, 1997; Draucker, Martsolf, Ross, & Rusk, 2007). The data collection proceeded in an iterative process as the study continued (Corbin & Strauss, 2008). Each interview, once completed, was immediately transcribed, coded, and analyzed, which helped refine and adjust the questions used in the next interview. The data collection process stopped at the saturation point when the information provided by participants became repetitive and each additional interviewee added little to the data (Corbin & Strauss, 2008). A total of 50 interviews were conducted. Thirty students and 20 parents of these 30 students were interviewed. Although 20 parents and students are paired, they were interviewed separately, with the other part of the family absent during the interviews. Each interview lasted about 40 to 60 minutes. All the interviews were audio-taped. Researchers also took written notes during the interviews. A pre-designed outline of interview questions was developed based on the literature and the research objectives of this study. The semi-structured interview questions for students included three categories of questions:

(1) Basic profile information of students including age, gender, city of origin, destination, length of the trip and accommodation. The sample questions are: 'can you briefly describe your trip itinerary?', 'how long was your trip?' and 'where did you stay during the trip?'
(2) Overseas study tour activities that the students participated in. The sample question is: 'what were the primary activities in your overseas study tour?'
(3) Reasons for students to participate in an overseas study tour. The sample question is: 'why did you want to attend an overseas study tour?'

The interview questions for parents primarily focused on the reasons for supporting their child on an overseas study tour. The sample question is: 'why would you want to support your child to participate in an overseas study tour?'

Data analysis

The analysis of interview data began with the completion of the first interview and lasted throughout the entire data collection process. All the interviews were conducted in Mandarin Chinese. They were first transcribed in Chinese and then translated into English by the second author and reviewed by the first author, both of whom are proficient in these two languages. The entire student data and the parents' data were merged and analyzed. After exacting the themes and subthemes from both sides, a comparative checking was conducted across a student's and his/her parents' transcripts. The basic structure of the coding scheme was developed based upon the first three research questions of this study: (1) Chinese adolescents' overseas study tour activities; (2) Chinese adolescents' reasons for participating in an overseas study tour; (3) parents' reasons for supporting the child's overseas study tour. In the initial round of data coding, researchers carefully read through the transcripts and identified themes under each coding scheme, respectively. The second round of data coding was conducted to examine the emerging categories and sub-categories under the above-mentioned coding scheme and to make connections between categories/sub-categories (Corbin & Strauss, 2008). For instance, three primary types of activities in overseas study tour were identified as: *sightseeing tours, schooling programs* and *socializing with the locals.* The final round of analysis moved from description, categorization and preliminary analysis toward interpretation. Throughout the entire procedure of data analysis, each transcript was carefully read multiple times to ensure that all the relevant data units were identified and synthesized. For example, for globalization: integration versus isolation was synthesized based on the subcategories from adolescents and parents.

Findings

The findings are delineated in five parts: (1) profile of informants; (2) activities on overseas study tours; (3) why overseas study tour? Chinese adolescents' perspective; (4) why overseas study tour? The parents' perspectives; (5) external forces that influence an overseas study tour. The results of data analysis present three major types of overseas study tour activities, five primary reasons for an overseas study tour from the Chinese adolescents' perspective and three primary reasons for supporting the child's overseas study tour from the parents' perspectives. Each major category with its subcategories is elaborated below.

Profile of informants

The profiles of Chinese adolescents and their parents are presented in Table 1. All of the informants' identities were disguised by using pseudonyms. A total of 30 Chinese adolescents who attended overseas study tours in the previous three years were interviewed. Twenty-three were female and seven were male, ranging from 13 to 19 years old. Parents of 20 students completed the interview, including 12 mothers and eight fathers. All of the informants were from big cities in China, including Shanghai, Beijing, Guangzhou, Nanjing, Wuhan, Chengdu and Shenzhen. The majority of the informants chose the United States as an overseas study tour destination (23 counts), followed by the United Kingdom (five counts), Japan (two counts) and Australia (one count). In general, the duration of an overseas study tour lasted about 15 to 40 days. Four types of accommodation were reported: home stay combined with hotel (12 counts), home stay (six counts), hotel (six counts) and university dormitory or others (six counts).

Activities in overseas study tour

Results of data analysis suggested three primary types of activities in an overseas study tour: *sightseeing tours, schooling programs* and *socializing with the locals*. Different types of these activities and examples drawn from Chinese adolescents' interviews are presented below and are summarized in Table 2.

Sightseeing tours

Two major types of sightseeing activities were reported by all student informants: (1) campus tour of prestigious universities, such as Ivy Leagues in the US and University of Oxford in the UK; (2) tour of well-known tourist attractions in destinations, such as Windsor Castle and the British Museum in London, the Louvre in France, etc.

Table 1. Profile of Chinese adolescents and their parents.

Student	Age	Gender	Parent	Age	Origin city	Destination country	Accommodation	Duration (days)
Abram	18	M	\		Nanjing	UK	Hotel + homestay	15
Adam	16	M	\		Chengdu	US	Homestay	30
Helena	14	F	\		Beijing	US	Hotel	25
Alan	16	M	Mother	45	Shanghai	US	Hotel + homestay	22
Hilary	17	F	\		Shanghai	US	Homestay	15
Alex	17	M	\		Shanghai	UK	University dormitory	20
Lea	14	F	Mother	44	Wuhan	UK	Hotel + homestay	30
Lina	16	F	Mother	45	Beijing	US	Homestay	40
Lily	19	F	Father	47	Guangzhou	Japan	University dormitory	12
Lisa	16	F	Mother	43	Wuhan	US	University dormitory	20
Ida	18	F	Mother	46	Nanjing	US	University dormitory	15
Albert	19	M	\		Shanghai	Japan	University reception	30
Ingrid	15	F	Mother	44	Wuhan	US	Homestay	20
Iris	14	F	Mother	39	Wuhan	UK	Homestay	20
Irma	19	F	Father	47	Nanjing	US	Stay with relatives	20
Jane	15	F	Mother	40	Wuhan	US; Australia	Hotel + homestay	20; 20
Janet	16	F	\		Wuhan	US	Hotel + homestay	18
Janice	16	F	Father	45	Wuhan	US	Hotel + homestay	20
Jolie	17	F	\		Wuhan	US	Hotel + homestay	25
Joyce	15	F	Father	44	Guangzhou	US	Hotel + homestay	21
Judy	14	F	\		Guangzhou	US	Hotel + homestay	21
Kathy	16	F	\		Shenzhen	US	Hotel + homestay	20
Karla	18	F	Father	44	Nanjing	US	Hotel	40
Kassie	17	F	Father	46	Wuhan	US	Hotel + homestay	22
Katherine	18	F	Father	48	Nanjing	US	Hotel	30
Kiley	14	F	Mother	49	Shenzhen	UK	Homestay	15
Kris	19	F	Father	44	Nanjing	US	Hotel	15
Alvin	16	M	Mother	48	Shenzhen	US	Hotel	15
Laura	14	F	Mother	44	Shenzhen	US	Hotel + homestay	30
Andrew	15	M	Mother	42	Shenzhen	US	Hotel	21

Table 2. Activities in overseas study tours.

Activities	Different types of activities
Sightseeing tours	Campus tour of prestigious universities Tour of well-known touristic attractions
Schooling programs	• Academic activities • Introduction of college education • Language learning • Regular classes at local schools Non-academic activities, such as painting, music and sports etc.
Socializing with the locals	Living with local families and participation in leisure activities, such as watching sports, parties, and fieldtrips etc.

Schooling programs

Schooling programs are the key component of overseas study tours, and refer to short-term study programs (1 to 2 weeks) in a local high school or university. Generally speaking, student informants reported three major types of academic activities and some non-academic activities in schooling programs. The first type is specifically designed for international students. These activities are offered by local universities, and aim to introduce the college education system to international students and help them prepare for future college applications in foreign countries. The second type of academic activity focuses on communication in foreign languages, such as writing, public speaking, etc. The third type pertains to the real schooling experience in foreign countries, which allows international students to join regular classes in local schools. In addition to the academic activities, some non-academic activities in schooling programs were reported by informants, such as painting, music, drama, sports, etc.

Socializing with the locals

The majority of Chinese adolescents lived with local families during their overseas study tour. Such an arrangement allowed them to socialize with local people through participating in different activities, such as watching sports games, participating in parties and attending religious events. Some host families even took them on weekend trips with family and friends. These social interactions with the locals provided them with first-hand experiences to help with understanding the culture, tradition and lifestyle in the foreign countries.

Why overseas study tour? Chinese adolescents' perspectives

The third theme focuses on why Chinese adolescents want to participate in overseas study tours. Five major reasons for an overseas study tour emerged from the data. They are: a taste of college education in foreign countries; global perspectives; socialization; independence; and novelty and fun.

A taste of college education in foreign countries

The majority of the adolescents expressed their desire to attend colleges in foreign countries and many have made the decision to study abroad. Therefore, this trip helps prepare them for college life in the near future. During the trip, these students were particularly interested in learning about majors, the campus environment, future job opportunities and the overall living conditions of international students. The following quote reflects the desire of this group of students:

> I have decided to attend college in Japan. I wanted to know more about the college before the upcoming school year. I was looking to learn about academic environment, placement of graduates and how Chinese are perceived by local people in Japan. Most importantly, I wanted to learn about Japanese culture and then I could decide whether I would be living there for a long time. (Albert, 19 years old, visited Japan)

There are also alternative examples. The following quote from Hilary reports a different purpose for taking an overseas study tour to the United States.

> My high school teacher really encouraged me to go study abroad in the US. However, I wasn't ready at that time. Psychologically, I was very dependent on my parents. So I was encouraged to take the tour to the American universities. I thought it might be a good chance for exploration. My parents would like me to go as well. If I want to study abroad for college, this tour can tell me whether it is a good choice for me. (Hilary, 17 years old, visited the United States)

In fact, the overseas study tour can serve as an opportunity for students to explore college education and life in foreign countries. In particular, when students have not decided to study abroad while parents or teachers are pushing them toward that decision, this trip can be helpful for all involved parties to address the conflicting ideas.

Global perspectives

Many Chinese adolescents expressed their desire to develop global perspectives through overseas study tours. Recognizing the challenges and opportunities of globalization, these Chinese adolescents are very eager to learn about different cultures and people in the world. To that end, an overseas study tour is attractive to many students as a great opportunity to develop global perspectives through the cross-cultural experience. The quote from Ida describes this sentiment:

> I want to have global perspectives. It means that I need to understand different cultures, learn about other countries and understand the differences between China and other countries. For example, I'm very interested in politics. I have heard and read so much about democracy. I want to know how Americans adopt democracy, what democracy really means and why it is a good thing. (Ida, 18 years old, visited the United States)

Gaining global perspectives is also reflected on students' intention of learning and improving English via an overseas study tour. Many informants reported their interest in English communications with host families and professors on school courses offered by the overseas study tour. The following quote of Jolie reveals this aspect:

> We have been learning English for so many years. Finally, it is time to use it. We stay in host family, take courses in school, do shopping, etc. Immersed in such an environment, we get to speak English every day. I want to improve my oral English and be able to communicate with native speakers. (Jolie, 17 years old, visited the United States)

Socialization

Socialization has been an important reason for people to travel. It is not an exception for Chinese adolescents to attend overseas study tours for this reason.

> I wanted to learn how to communicate with Americans and make friends with them. Since it was a group tour, I need to learn how to get along with other group members. I think it is important to learn how to socialize … and how to communicate with other people in a new environment. (Janet, 16 years old, visited the United States)

As reported by Janet, many students mentioned two aspects of socialization in an overseas study tour: (1) to meet new friends from the host universities and host families; (2) to help strengthen the friendship with current classmates and friends.

Independence

Several adolescents indicated that they can obtain a sense of independence via an overseas study tour. Alvin describes this sentiment as follows:

> I wanted to go 'outside' by my own. My parents always traveled with me. This time, I can go with my classmates. It is an opportunity for me to gain independence. (Alvin, 16 years old, visited the United States)

Being often the only child in the family, most adolescents acknowledged their lack of independence at home. An overseas study tour was the first time for them to travel without parents.

Novelty and fun

In addition to study opportunities, another important reason for Chinese adolescents to attend an overseas study tour is to seek novelty and fun. Several students mentioned that the destinations of overseas study tours were attractive to them because they had never been to those places. In particular, the students who do not have plans for study abroad are more excited about the sightseeing and recreational activities offered by the tour.

> I'm very interested in the United Kingdom. Particularly, I like British culture, such as the British lifestyle, architecture and history. I have been collecting souvenirs and information about Britain since a long time ago. So when my mom told me about this trip, I immediately signed up. (Alex, 17 years old, visited the United Kingdom)
> I was not planning to study abroad, so I wasn't so focused on a campus tour. I love travel. My main purpose was to visit interesting places and try delicious food. I wanted to visit the Disney World. (Joyce, 15 years old, visited the United States)

Why overseas study tour? The parents' perspectives

The data analysis result reveals three primary reasons that Chinese parents are willing to support their child on an overseas study tour. They are: the child's future plans; child's social development; and lost college dreams. Each major category, with examples, is illustrated below.

Child's future plans

College education. It has been constantly mentioned by Chinese parents that an overseas study tour helps with the child's future education plans no matter whether or not the child has decided to study abroad. Ingrid has decided to study abroad in college. From her mother's perspective,

> My daughter has determined to study abroad. I wanted her to know about the condition of international students in foreign countries. Then she will have some preparation ahead of time. She would know whether this [study abroad] is the life that she wants to have in the next few years. (Ingrid's mother, 40 years old)

It is believed that this tour allows Ingrid to gain more understandings of the foreign universities and her future condition as an international student there. The child can prepare herself in advance.

On the other hand, some students, such as Laura, have not decided about studying abroad. As said by her mother,

> Our initial thought is that my daughter can have a clear goal after coming back [from the overseas study tour]. We hope she will study abroad for college. She needs to know what type of college she wants to attend in the future. She needs to make a plan for herself and decide whether study abroad fits her. She is expected to have a clear picture of all these things and be more motivated to study for college when she comes back [from the overseas study tour]. (Laura's mother, 44 years old)

Laura's mom expects she can have a better idea about the future study abroad as an international student. Parents would like to give more empowerment to the child for her future education so she can plan it accordingly.

Post-college development. In addition to college education, many parents expressed their concerns with their child's post-college development. For example, Lisa's mother commented as follows:

> In my view, what Chinese high school students lack most is a global perspective. Three years of high school are the golden time for kids' development. If they are forced to only focus on test scores, they are missing out a lot of important things. It is tunnel vision to let kids just focus on scores. After finishing college education, students are evaluated not only by textbook knowledge but also by their experience, insight and vision. If a student has a lot of travel experiences and social knowledge, she will be more competitive. (Lisa's mother, 43 years old)

Parents are concerned with Chinese school education that lacks a global perspectives. On the other hand, they noted the importance of global perspectives in enhancing the child's competitiveness in the future society. They believe an overseas study tour can provide the child with opportunities to travel globally, understanding different cultures and broadening their views. Thus, the gaps in China's current education can be filled via such a tour.

Child's personal development

Many parents believe that an overseas study tour improves their children's personal development. Given that this is the first time for most adolescents to have an overseas trip not accompanied by parents, such an experience is expected to provide learning opportunities to them, such as adapting to the new environment, being independent and mature, making new friends and so on. The following quote from Ingrid's mother reflects this point of view:

> I hope she can make more friends while traveling with other students. It is important for her to learn how to socialize with others. The single-child generation particularly lacks such skills. I am actually 'observing' whether my daughter has a collective consciousness and whether she can get along with the host family. (Ingrid's mother, 44 years old)

In addition, Kiley's mother said,

> I think it [the overseas study tour] is a very good opportunity for her to learn how to adapt to a new environment. Be away from parents, stay in a new place with classmates, speak the new language and learn how to handle things independently … these are all good training for her. Although it's only about one month, she would become more mature and get to know how to think from other people's perspectives. (Kiley's mother, 49 years old)

Apparently, parents appreciated particular features of an overseas study tour, which are that students will live with host families, meet local people, experience a new environment, etc. They believe that such experiences can be beneficial to their child, who grew up with so much care and attentions as the single child in the family. They also implied in the interviews that their children lack independence and hope the tour can help them develop their independence skills and become more mature.

The lost college dream

Most parents that participated in this study were born in the 1960s and their juvenile years suffered insufficient education resources. They went through China's decade-long Culture Revolution (1966–1976), when the education system in the whole country was completely destroyed. China's national college entrance examination was cancelled from 1966 until 1977. Several parents mentioned that they did not have a good opportunity for higher education when they were young. Studying abroad was a aspiration they could not reach at that time. It was their hope, deep in their heart, that their child could fulfill their lost dreams of college education. Kiley's mother described this sentiment:

> Study abroad has been my husband's dream since his college years. Our generation has many lost dreams. We don't want to admit that we want our kid to realize our dreams, but we are doing it unconsciously. In the 1990s, I had a business trip to the US. I found that the US was so much more advanced than China in many aspects. It is already impossible for me to go abroad for college education. So I want my child to go abroad and see a different world. (Kiley's mother, 49 years old)

Apparently, from the perspective of Kiley's mother, she and her husband tend to push their child to pursue the opportunities they did not have when they were Kiley's age, regardless of the child's willingness. A similar idea has been mentioned by several parents in the interviews.

External forces that influence an overseas study tour

In-depth analysis of the interviews results in a number of forces from the external environment that shape Chinese adolescents and their parents' perspectives on overseas study tours. For example, the process of globalization and China's socio-economic modernization is the fundamental background that deeply influences Chinese society and its people. In addition, Chinese traditional cultural values and China's contemporary single-child family structure both play an important role in reshaping the parents' and adolescents' understandings of overseas study tours. In fact, a closer look into these forces leads to a further understandings of contemporary Chinese middle-class, regarding their beliefs, values, concerns, etc.

Globalization: integration versus isolation

Globalization is flattening the world and its impacts have been felt far and wide. Today's world is more closely interconnected than ever before due to technological advancements, a breakdown of geographical borders and more convenient means of communication and transportation, etc. (Hirst, Thompson, & Bromley, 2009). The process of globalization brings enormous opportunities to organizations and people from different countries to exchange and communicate with each other. In the last few decades, China's socio-economic modernization has been deeply associated with liberation and globalization. Since 1978, China has adopted a reform and opening-up policy (Wall, 1993) and gradually transformed itself from a mandatory planning economy to a modern market economy (Logan, 2011). China's increasing integration into the world by adopting a market-oriented economic reform has led to fast economic growth over the past three decades. In addition to the socio-economic development, Chinese society has been extensively influenced by the process of globalization. With the door open, Chinese society is experiencing new and different beliefs, values and cultures, which greatly affect Chinese people's lives and ways of thinking. In addition, an increasing number of Chinese have the opportunity to travel abroad and to interact with cultures other than their own. Through the above-mentioned various cross-cultural communications, more and more Chinese people have opened up their minds, understood different beliefs, values and cultures, developed their own understandings of the world and progressively established closer contacts with other parts of the world.

During the process of China's socio-economic reforms and continued integration into the global system, a new middle class has been rapidly growing in China (Farrell, Gersch, & Stephenson, 2006). The people who have newly moved into this social class are the beneficiaries of China's market-oriented economic agenda over the past three decades. Compared with most people in China, the new middle class have gained higher income and social status. Undoubtedly, the economic development laid the materialistic foundation for the emergence and growth of China's middle class. Additionally, the social transformation from an extreme centralism to a socialist democratic political system since the 1980s provides favorable conditions for the prosperity of China's middle class (Zhou, 2008). Different from most developed countries whose middle class people account for about 50% of the whole population and represent the mainstream of a society, the new middle class in China constitutes less than 20% of the whole population and is regarded as the elite stratum of the society (Zhou, 2008). This social group stands out as a salient force in the process of China's socio-economic modernization (Chen & Lu, 2011). Most people in this social stratum are well-educated and have specialized skills. Their occupations mostly fall into the categories that include government officials or employees, administrators or managers in enterprises and high-tech professionals (Lu, 2002). Accordingly, the life experiences associated with these types of professions allow the middle class to have more exposure to western values and easier access to western societies (Tang, Woods, & Zhao, 2009).

As reflected in this study, the majority of the patrons of overseas study tours are from the middle-class and upper-middle class families in China's big cities. The parents represent the social stratum of China's new middle class that has emerged from China's socio-economic modernization over the past three decades. These middle-class and upper-middle class adults are among the first group of people in Chinese society who witnessed globalization in China and opened up their minds toward its benefits and challenges. Their life experiences lead them to recognize that China has become deeply connected with other parts of world and that having a global mindset is essential for individuals to be successful in today's China. In addition, they tend to show a great degree of openness to cultural diversity and cross-cultural communications. With such values in mind, these parents tend to guide their children and equip them with a global mindset in addition to the knowledge and skills for future development. They deeply believe that openness to cultural diversity and understanding different views from other societies are very much necessary for preparing their child to be successful in today's more connected world community. The following quote from Jane's mother has presented such an emphasis on equipping the next generation with a global mindset and openness to cultural diversity.

> We need to nurture the kids to have a 'global mindset' and to understand what people in other countries are doing. The gap between one country and another is getting smaller. The US is not better than us in all aspects. In some aspects, they are better than us ... I think the China society is much more open than ten years or twenty years ago, and more people are traveling abroad and studying abroad to learn 'advanced experience' from developed countries. (Jane's mother, 40 years old)

On the one hand, China and Chinese middle-class have been eager to integrate into the world's economic development under the globalization context. On the other hand, some parts in Chinese society are staying isolated from globalization, such as the education system, the political system, etc. For example, both adolescents and their parents have mentioned that current Chinese school education has too much emphasis on academic scores and overlooks other aspects of youth development. Lea's mother expressed her concern with the pure emphasis on scores and exams of Chinese schools,

> China's school education is way too much focused on exams and scores. What students are learning at school is very limited. It [the overseas study tour] is a great opportunity for the child to enrich her experience, open up the mind, as well as improve language skills. They are more advanced than us in education and in many other aspects. It's a chance for my daughter to take a look. (Lea's mother, 44 years old)

In addition, some informants have talked about the different political system of China compared with many developed countries in the world. They found that China has made great progress in terms of economic openness in the past decades but not much in its political system. The following quote from Lisa's mother presents this opinion.

> We have a lot of things to learn from the United States, such as democracy, human rights, economics, values, etc. My travel experience in the United States is that American society is more diverse, not only in race, but also in values. They are more tolerant to different values. In China, we are more likely to use the 'mainstream criteria' to judge everything, while neglecting other different voices. (Lisa's mother, 43 years old)

Apparently, parents who represent Chinese middle-class have recognized the importance of global perspectives in today's world. They are concerned that this piece is missing in China's school education for their children. An overseas study tour provides a good supplement and addresses their concerns. They expect their children can benefit from this tour and understand the globalization via such personal experiences.

Fathers and sons: obedience versus independence

This study investigates both Chinese adolescents and their parents' perspectives on overseas study tours. The perspectives of two generations reflect an interesting and dynamic interaction between parents and children in contemporary Chinese society. It has been well documented in literature that parents, as the guardians of their children, have a higher status in the Chinese traditional family structure (Bond & Hwang, 1986). One can find such hierarchy roots in Confucian ideology, in that the child is expected to be obedient and respectful; parents are the authoritative in the family and are responsible for making important decisions for the child (Xu et al., 2005). Interestingly, the findings of this study present the evolution of the parent–child relationship as reflected in the decision-making process of parents and adolescents, which can be attributed to the cultural clash between Chinese traditional culture and western culture as well as the one-child family structure in today's China society. On the one hand, the findings of this study indicate that the reasons why the child attends overseas study tours largely overlap with the reasons why the parents support the child for the tour. For instance, 'a taste of college education in developed countries', 'global perspective' and 'socialization' are mentioned by the child and parents. The 'overlapping' implies that parents may have exerted a considerable amount of influences on shaping the child's perspectives regarding overseas study tours. For instance, the following quote from Adam indicates how his parents influenced his perspectives of and decisions on an overseas study tour.

> Initially I didn't want to participate, because I didn't want to waste one month on travel. I would be so falling behind on my school work. However, my parents both persuaded me to take this chance. They told me that it would be a totally new experience which would have a major impact on my future life. It is importance to go outside, broaden my view and enrich my life experience. Well, then I gave it a try. (Adam, 16 years old, visited the United States)

Although parents tend to influence the child's perspectives on an overseas study tour, they did not play an absolutely dominant role in the decision-making process. The majority of the adolescents mentioned that although their parents may have initiated the idea and suggested they attend overseas study tours, to a large extent the adolescents can make the final decision independently. This may appear to contradict the traditional Chinese cultural values that the child has to obey the parents. However, the parent–child relationship in today's transitional China society features the paradox of obedience and independence. Both sides (parents and adolescents) have mentioned 'independence' as one reason for an overseas study tour: parents expect the child can learn to be independent and adolescents look forward to the sense of independence via the experience. Such a paradox and the emergence of independence in the parent–child relationship can be attributed to several factors in today's Chinese society. The first important factor is the cultural clash between Chinese traditional culture and western culture resulting from the modernization and westernization over the past several decades in China (Faure & Fang, 2008). To some degree, these social and economic changes have not only stratified Chinese society (e.g. led to the growth of the middle class) but also transformed parents' attitudes toward child-rearing styles. Parents in this study are well-educated middle class or upper-middle class, who are familiar with Chinese traditional culture and largely exposed to western culture values. As such, they tend to give the child more freedom and independence in decision-making with regard to participating in an overseas study tour. For instance, Ingrid's following quote illustrates such liberal parenting style in her family.

> My family is not a traditional Chinese family. My parents give me as much freedom as I need. They often tell me that, they won't force me to do anything, choose a career, or plan the future for me. It's all up to me. But whenever I need them, they will always be there for me, and back me up. My parents are very liberal. (Ingrid, 15 years old, visited the United States)

Another important factor to explain the paradox of obedience and independence in the parent–child relationship is the single-child family structure in today's China. The Chinese government implemented the single-child policy from 1978 to 2015. This almost three-decade long policy has resulted in a 'child-centered' phenomenon in today's Chinese families, especially among well-educated and high-income populations (Xu et al., 2005). Within this single-child family structure, parents tend to provide more freedom and flexibility to their only child, who is usually regarded as the center and future of the whole family. Accordingly, the child has more power in the decision-making process. The following quote from Katherine's father reflects this aspect.

> Parents' opinions are only one aspect. We focus more on how the kid thinks about study abroad and what is her opinion. Even though parents have high expectations on the child, but what if the child doesn't study hard. The final results may turn out not to be good. So, no matter it's about study abroad or overseas study tour, my daughter has been constantly discussing it with her mom. Taking account of my daughter's opinion, the final decision is made upon the whole family's agreement. (Katherine's father, 48 years old)

In summary, the parent–child relationship in China is changing to become more equal, friend-like and independent due to cultural clashes between Chinese traditional culture and western culture, as well as China's single child family structure. Parents are willing to grant more power to the child in matters that relate to the child.

The dream: lost versus resuming

In addition, this study finds that when Chinese parents communicate their opinions on overseas study tours with their child, they tend to extend or even pass on their lost college dreams to the next generation. Interestingly, it seems to be well received by the child informants that they should resume their parents' lost dreams of overseas education. For example, Kiley mentions such interaction with her parents in the following quote:

> My parents always want me to study abroad. My dad told me many times that one of his dreams was to study in the United States when he was young. I agree with them that foreign universities are very good, and I also want to study abroad one day. Although I didn't go to the United States, I went to London for the overseas study tour because I'm more interested in British schools. My parents are still very happy about it. (Kiley, 14 years old, visited the United Kingdom)

Over thousands of years in China society, education attainments have constantly been considered as an extremely important means for an individual to ascend the social ladder (Yu & Suen, 2005). In the classical book of China education *Analects of Confucius*, Confucius said: 'Excellent learning, a prosperous official career achieved' [学而优则仕]. This viewpoint has greatly influenced Chinese generations and they also believe 'Only to be a scholar is being the top of the society' [万般皆下品，惟有读书高]. The importance of college education still holds true in the society. College education remains the primary path of upward mobility for most of people and a symbol of cultural/social capital (Lin & Bian, 1991; Yu & Suen, 2005). The majority of Chinese adolescents' parents nowadays experienced the Cultural Revolution (1966–1976) in their youth. These parents did not have opportunities for higher education abroad or even domestically at that time. Such a tragedy of the times has left their lives with regrets and unfulfilled dreams for better education and more opportunities. Such regrets result in stronger hopes due to the prevalent perceptions among the Chinese middle-class that schools in developed countries are more prestigious and so could offer better study conditions than domestic ones. With the hierarchical parent–child relationship defined in Chinese traditional culture, the parents in elite groups tend to use their authority and higher status in the relationship to impose their unfulfilled ideals and dreams on children or successors. On the other hand, filial piety [孝], as a key virtue in Chinese culture denoting that one should love, respect and support one's parents, is deeply believed by Chinese adolescents nowadays and leads them to have the intention of fulfilling the ideals and dreams that their parents have being unable to accomplish.

Conclusions and discussions

Recognizing that tourism can reflect the economic, social and cultural forces in a society, this study uses Chinese adolescents' overseas study tours to elaborate on the interactions between tourism and economic, social and cultural forces in Chinese society. Both Chinese adolescents and their parents' perspectives have been taken account to discover multiple reasons for overseas study tours. Furthermore, three primary forces were identified from the interviews with parents and adolescents.

This study makes contributions to existing consumer behavior research in the tourism area. It includes forces from the macro environment in the analysis and understanding of people's decision-making process. Nowadays, tourism has become a significant proportion of more and more people's lives. In addition to being an economic activity, tourism exerts social influences on people, including destination residents and tourists. Given that tourists are individuals in their society, the forces related to economic, social, cultural, and political aspects in their society have impacts on individuals, including their travel decision making and behavior. On the other hand, their tourism behavior can reflect and demonstrate those forces in society. This study recognizes the changing role of tourism in today's fast-changing China: it is 'not only an outcome of change, but a catalyst for further change' in China (Ryan & Huang, 2013, p. xix). Taking Chinese adolescents' overseas study tours as an example, the current study has explored the influencing forces from the external environment, which are also important forces changing the Chinese middle-class. This study sheds new light on tourist behavior research on answering the 'why' questions with root causes and in-depth analysis of tourists' external environment in the society they live in.

Furthermore, the current research calls for further investigations into Chinese outbound tourists. In particular, two issues should be investigated given that more and more Chinese tourists travel abroad. First, the growing number of Chinese tourists has led to multi-market segments based on the benefits sought, demographic information, travel purpose and behavior, etc. However, extant research on Chinese outbound tourists still takes the 'mass market' perspective, viewing them as a homogeneous group. As many countries in the world are looking forward to welcoming a big wave of Chinese tourists in the upcoming years, it is urgent that they understand the travel behaviors of Chinese tourists in different segments in terms of their unique socio-cultural and psychological accounts. The current study is conducted to initiate this line of research, investigating Chinese adolescents' overseas study tours to developed countries. Second, as more and more Chinese tourists plan to travel to countries that are significantly culturally distant from China, these people are facing cross-cultural challenges in their travel. This study notes and examines the impacts of forces from their original society on their travel decision making. This accords with the mobilities theories that state that people's daily lives and travel are linked in many ways. This study brings academic attention to this issue of the blurred border between people's daily life and travel.

This study provides several important practical implications for travel agencies and destination marketers whose target market is Chinese adolescents. First, it is anticipated that the volume of overseas study tours among Chinese adolescents will continue to grow, providing more opportunities for tourism businesses in both China and destination countries. The findings of this study can help travel agencies develop appropriate products to cater to this fast-growing market segment. For instance, getting a taste of college education in developed countries is identified as a major reason for Chinese adolescents to attend overseas study tours. As such, travel agencies can design more diverse campus activities which can add to Chinese adolescents' educational experiences in a destination country. For example, face-to-face communication with local students, classroom experience, college sport activities, living in dorms and so on, can be added to the tour itinerary for Chinese adolescents to get a better taste of college life in the developed countries.

Another important suggestion for travel agencies is to develop slightly different marketing communications with Chinese adolescents and their parents. The findings of this study indicate two reasons for this suggestion. First, it is found that parents exert a significant amount of influence on the child's perspectives on, and the decision-making with regard to, an overseas study tour. Thus, understanding the parents' perspectives enables travel agencies to better design their marketing strategies to attract both the children and their parents to capitalize on this lucrative segment. Second, the findings of this study also indicate the different reasons for an overseas study tour from parents and adolescents. The parents' perspectives are largely centered on preparing for the child's future college education and post-college development, to supplement the current Chinese school education. The children's perspectives are more scattered, such as independence, a global perspective, socialization and fun. As such, the marketing communication toward parents should be focused more on the educational benefits that their child cannot receive from Chinese school education. Marketing communications toward children should be more focused on socialization, fun, independence from parents, different cultures and new people, and so on.

In addition, this study helps destination marketers in developed countries, such as the UK, US, etc., to enrich their market portfolios of Chinese tourists in the future. In particular, it provides implications and opportunities for college towns to attract more Chinese tourists and adolescents who can be their prospective students. As an important market segment of China's outbound tourism, overseas educational tours for Chinese adolescents will continue to expand (Nan, 2014). Yet, the understanding of this market segment remains limited. Findings of this study, such as the adolescents' activities, reasons for the tour, etc., can help DMOs of college towns to design the appropriate tourism product in order to attract this growing market segment from China.

This study is not without limitations. These limitations, however, also indicate opportunities for future research. The current study identified forces from the external environment and elaborated on the influences of these forces in narratives. However, the extent/degree of the impacts of these forces on the decision-making process cannot be examined by qualitative interviews. In addition, the importance of the different reasons between parents' and adolescents' perspectives cannot be clearly presented by the qualitative data. Future research is suggested to use quantitative methods with a larger sample size to measure the impacts of forces and the importance of the reasons for an overseas study tour. Future studies are also encouraged to investigate Chinese adolescents' experiences in the host destination and the transformative impacts on them after the trip.

Disclosure statement

No potential conflict of interest was reported by the authors.

Funding

This study is not supported by any finding agency.

References

Anderson, P. H., Lawton, L., Rexeisen, R. J., & Hubbard, A. C. (2006). Short-term study abroad and intercultural sensitivity: A pilot study. *International Journal of Intercultural Relations, 30* (4), 457–469.

Behnke, C., Seo, S., & Miller, K. (2014). Enhancing the study abroad experience: A longitudinal analysis of hospitality-oriented, study abroad program evaluations. *Tourism Management, 42*, 271–281.

Bond, M. H., & Hwang, K. K. (1986). *The social psychology of Chinese people.* In M. H. Bond (Ed.), The psychology of the Chinese people (pp. 213-266). New York, NY: Oxford University Press.

Bosselman, R. H., Fernsten, J. A., Manning, P. B., & Kisseleff, M. (1989). The international study abroad experience and its effects on hospitality students. *Journal of Hospitality & Tourism Research, 13*(3), 287–297.

Carlson, L. E. S., & Grossbart, S. (1988). Parental of style and consumer socialization children. *Journal of Consumer Research, 15*(1), 77–94.

Chan, K., & McNeal, J. U. (2003). Parent-child communications about consumption and advertising in China. *Journal of Consumer Marketing, 20*(4), 317–334.

Chao, R. K. (1996). Chinese and European American mothers' beliefs about the role of parenting in children's school success. *Journal of Cross-Cultural Psychology, 27*(4), 403–423.

Chen, J., & Lu, C. (2011). Democratization and the middle class in China. The middle class's attitudes toward democracy. *Political Research Quarterly, 64*(3), 705–719.

Chieffo, L., & Griffiths, L. (2009). Here to stay: Increasing acceptance of short-term study abroad programs. In R. Lewin (Ed.), *The handbook of practice and research in study abroad: Higher education and the quest for global citizenship* (pp. 365–380). New York, NY: Routledge.
Corbin, J., & Strauss, A. (2008). *Basics of qualitative research: Techniques and procedures for developing grounded theory* (3nd ed.). Thousand Oaks, CA: Sage Publications.
Coyne, I. T. (1997). Sampling in qualitative research. Purposeful and theoretical sampling; merging or clear boundaries? *Journal of Advanced Nursing*, *26*(3), 623–630.
Doyle, S., Gendall, P., Meyer, L. H., Hoek, J., Tait, C., McKenzie, L., & Loorparg, A. (2010). An investigation of factors associated with student participation in study abroad. *Journal of Studies in International Education*, *14*(5), 471–490.
Draucker, C. B., Martsolf, D. S., Ross, R., & Rusk, T. B. (2007). Theoretical sampling and category development in grounded theory. *Qualitative Health Research*, *17*(8), 1137–1148.
Dukes, R., Lockwood, E., Oliver, H., Pezalla, C., & Wilker, C. (1994). A longitudinal study of a semester at sea voyage. *Annals of Tourism Research*, *21*(3), 489–498.
Farrell, D., Gersch, U. A., & Stephenson, E. (2006). The value of China's emerging middle class. *McKinsey Quarterly*, *2*(I), 60.
Faure, G. O., & Fang, T. (2008). Changing Chinese values: Keeping up with paradoxes. *International Business Review*, *17*(2), 194–207.
Fu, X., Huang, Z., Li, Q., & Kirillova, K. (2017). Dissecting Chinese adolescents' overseas educational travel experiences: Movements, representations and practices. *Current Issues in Tourism*, 21(10), 1115-1136.
Hannam, K., Butler, G., & Paris, C. M. (2013). Developments and key issues in tourism mobilities. *Annals of Tourism Research*, *44*(1), 171–185.
Hirst, P., Thompson, G., & Bromley, S. (2009). *Globalization in question*. Cambridge, UK: Polity Press.
Houser, C., Brannstrom, C., Quiring, S. M., & Lemmons, K. K. (2011). Study abroad field trip improves test performance through engagement and new social networks. *Journal of Geography in Higher Education*, *35*(4), 513–528.
ICEF Monitor Report. (2014). *Summing up international student mobility in 2014*. Retrieved April 20, 2016, from http://monitor.icef.com/2014/02/summing-up-international-student-mobility-in-2014/
Juvan, E., & Lesjak, M. (2011). Erasmus exchange program: Opportunity for professional growth or sponsored vacations? *Journal of Hospitality & Tourism Education*, *23*(2), 23–29.
Lee, M. (2012). *The complete history of study abroad*. Retrieved May 5, 2018, from http://www.gooverseas.com/go-abroad-blog/history-study-abroad-part-1
Lin, N., & Bian, Y. (1991). Getting ahead in urban China. *American Journal of Sociology*, *97*(3), 657–688.
Logan, J. (Ed.). (2011). *The new Chinese city: Globalization and market reform* (Vol. 52). Oxford: Blackwell.
Lu, X. (2002). *当代中国生活阶层研究报告* [Report on contemporary Chinese social strata]. Beijing: 社会科学出版社 [Social Science Press].
Lyons, K., Hanley, J., Wearing, S., & Neil, J. (2012). Gap year volunteer tourism: Myths of global citizenship? *Annals of Tourism Research*, *39*(1), 361–378.
Moschis, G. P. (1985). The role consumer of family communication of socialization adolescents children. *Journal of Consumer Research*, *11*(4), 898–913.
Nan, J. (2014). *游学市场"太火爆" 各国游学费用盘点* [The edu2cational abroad travel is "too hot": Compare the cost to different countries]. *决胜网* [Juesheng wang]. Retrieved August 7, 2015, from http://news.juesheng.com/detail/100142.html
Nyaupane, G. P., Paris, C. M., & Teye, V. (2011). Study abroad motivations, destination selection and pre-trip attitude formation. *International Journal of Tourism Research*, *13*(3), 205–217.
Pickert, S. M. (1992). Preparing for a global community: Achieving an international perspective in higher education, ASHE-ERIC Higher Education Report No.2. Washington, DC: School of Education and Human Development, George Washington University.

Pizam, A., Jafari, J., & Milman, A. (1991). Influence of tourism on attitudes: US students visiting USSR. *Tourism Management, 12*(1), 47–54.

Qu, G. (2013), *海外修学游提前升温* [overseas study tours gets even hotter]. Retrieved August 7 2015 from: http://epaper.xiancn.com/xawb/html/2013-04/24/content_200705.htm

Redden, E. (2014). *International enrollment up*. Retrieved April 20, 2016, from https://www.insidehighered.com/news/2014/11/17/open-doors-report-finds-increases-international-enrollment-study-abroad

Ritchie, B. W. (2003). *Managing educational tourism*. Clevedon, UK: Channel View Publications.

Ryan, C., & Huang, S. (2013). *Tourism in China: Destinations, planning and experiences*. Clevedon, UK: Channel View Publications.

Sin, H. L. (2009). Volunteer tourism—"Involve me and I will learn". *Annals of Tourism Research, 36*(3), 480–501.

Tang, M., Woods, D., & Zhao, J. (2009). The attitudes of the Chinese middle class towards democracy. *Journal of Chinese Political Science, 14*(1), 81–95.

Thibadoux, G. M., & Helms, M. M. (1989). An analysis of a successful international study abroad program. *Annals of Tourism Research, 16*(4), 564–566.

Toncar, M. F., & Cudmore, B. V. (2000). The overseas internship experience. *Journal of Marketing Education, 22*(1), 54–63.

Van Hoof, H. B. (2000). The international internship as part of the hospitality management curriculum: Combining work experience with international exposure. *Journal of Hospitality & Tourism Education, 12*(1), 6–15.

van't Klooster, E., van Wijk, J., Go, F., & van Rekom, J. (2008). Educational travel: The overseas internship. *Annals of Tourism Research, 35*(3), 690–711.

Wall, D. (1993). China's economic reform and opening-up process: The role of the special economic zones. *Development Policy Review, 11*(3), 243–260.

Wearing, S. (2001). *Volunteer tourism: Experiences that make a difference*. Wallingford: CABI Publishing.

Xu, Y., Farver, J. A. M., Zhang, Z., Zeng, Q., Yu, L., & Cai, B. (2005). Mainland Chinese parenting styles and parent–Child interaction. *International Journal of Behavioral Development, 29*(6), 524–531.

Yan, A., & Gray, B. (1994). Bargaining power, management control and performance in United States – China joint ventures: A comparative case study. *The Academy of Management Journal, 37*(6), 1478–1517.

Yu, L., & Suen, H. K. (2005). Historical and contemporary exam-driven education fever in China. *KEDI Journal of Educational Policy, 2*(1), 17–33.

Zheng, X. (2015). *500,000 Chinese students attend summer overseas study tours* [暑期将至 50万中国学生出国游学]. Retrieved from http://www.epochtimes.com/gb/15/5/26/n4442852.htm

Zhou, X. (2008). Chinese Middle Class: Reality or Illusion, In C. Jaffrelot & P. Van der Veer (Eds.), *Patterns of Middle Class Consumption in India and China (pp. 110-126)*, New Delhi: Sage Publications.

Exploring the Influence of Family Holiday Travel on the Subjective Well-being of Chinese Adolescents

Mingjie Gao, Mark E. Havitz and Luke R. Potwarka

ABSTRACT

This study aimed to explore the influence of family holiday travel on the subjective well-being (SWB) of Chinese adolescents. Surveys were distributed at two public middle schools in the urban area of a large city located in the eastern part of Mainland China. Participants were middle school students aged between 12 and 15 years (grades 7–9). By using Labor Day in China as an experimental context, this study applied a longitudinal research design. Findings suggest that family holiday travel influences the global life satisfaction; contentment with school, self and leisure life; positive and negative affects of adolescents. In particular, there is a short-term lift-up effect of family holiday travel on the SWB of adolescent travelers. However, the results suggest that the benefits of family holiday travel in terms of SWB diminish when adolescent students return to school. Moreover, students who travel with their families during holidays have significantly higher post-holiday SWB than their non-traveling counterparts. The current study advances our knowledge on the influence of family travel on the SWB of adolescents. Recommendations for parents, schools, and the government were put forward.

家庭旅游对青少年主观幸福感影响的研究

摘要

本研究旨在探索家庭旅游对中国青少年主观幸福感的影响。研究对象为年龄介于12至15岁的初中学生。以劳动节小长假为实证研究的背景，本研究采用了纵向研究的设计方法，调查问卷于中国大陆东部地区某大城市的两所中学进行发放。研究发现家庭旅游能够提升青少年的整体生活满意度。此外，家庭旅游对于青少年的自我认可，对学校生活和休闲生活的满意度，以及对积极和消极情绪存在显著影响。具体来说，家庭旅游能够在短期内提高青少年旅游者的主观幸福感。然而，家庭旅游对于青少年主观幸福感的积极影响在青少年返校后有所削减。在假期中，和家人旅行的青少年相比没有旅行的青少年其主观幸福感显著较高。此研究对于加深理解家庭旅游和青少年主观幸福感的关系有一定贡献，文末分别针对家长、学校以及休假制度的完善提出了相关建议。

Introduction

For generations, people have always looked for ways to lead happy lives. According to research on the subjective well-being (SWB) of Chinese middle school students, the average levels of SWB among Chinese adolescents are not satisfied (Hu, Ma, Hu, Deng, & Mei, 2010). Research suggests that most stressors that decrease the SWB of Chinese adolescents are related to education (Hu et al., 2010; Tian, Liu, Huang, & Huebner, 2013). The low levels of SWB of Chinese adolescents deserve attention, and understanding how we can help adolescents buffer the negative influence of stressors and promote their SWB effectively are important. Tourism literature has suggested that traveling can be a beneficial way to improve the SWB of individuals (Chen & Petrick, 2013; Uysal, Sirgy, Woo, & Kim, 2016). This study aims to examine the influence of family holiday travel on the SWB of Chinese adolescents.

Generally, recreational travel increases post-travel well-being (de Bloom et al., 2011; Nawijn, 2011a; Nawijn, Marchand, Veenhoven, & Vingerhoets, 2010), and people who travel are happier than those who do not travel (Gilbert & Abdullah, 2004; Nawijn, 2011b). Many studies have explored the ways in which recreational travel boosts post-travel SWB. In particular, pleasant activities (de Bloom et al., 2011), recovery experiences (Fritz & Sonnentag, 2006; Nawijn et al., 2010) and trip satisfaction (Neal, Uysal, & Sirgy, 2007; Sirgy, Kruger, Lee, & Grace, 2011) are key factors that enable travel to contribute to the overall life satisfaction of travelers. Moreover, holiday stress (Nawijn, 2011a) and conflicts with companions during trips (Havitz, Shaw, & Delamere, 2010; Rosenblatt & Russell, 1975) are the most important determinants that decrease the life satisfaction of travelers.

Extant studies have also revealed a positive relationship between family travel and the family well-being (Durko & Petrick, 2013). Amid fluid family dynamics and influential factors in families, family travel has been recognized as an effective means for family members to improve communications within relationships, strengthen family functioning, and thus enhance family well-being (Lehto, Choi, Lin, & MacDermid, 2009; Lehto, Lin, Chen, & Choi, 2012). However, little attention has been paid to understand the role that family travel plays in producing beneficial outcomes from children's perspectives. Only a partial understanding of travel benefits can be gained if children's perspectives are not included (Small, 2008). In reality, children's voices have been included in the social study of childhood in which children are treated as active social actors (Seymour & McNamee, 2012); however, the same has yet to be observed in the tourism literature. Several studies have demonstrated that children's expectations and evaluations of family travel are sometimes different from their parents' perspectives (Carr, 2006; Fu, Lehto, & Park, 2014; Gram, 2005). Therefore, the benefits of travel for adults may not be fully applied to children. Understanding the benefits of family travel on SWB from children's perspectives is clearly of great importance.

The purpose of this study is to examine the influence of family holiday travel on the SWB of Chinese adolescents. In particular, this study attempted to answer three research questions: (1) Does family travel have a short-term effect on adolescent SWB? (2) Does family travel have a long-term effect on adolescent SWB? (3) Do differences in self-reported SWB exist between adolescents who travel and those who do not travel during family holidays?

Literature review

Conceptualization of SWB

According to the psychology literature, SWB is under the umbrella of quality of life (Diener & Suh, 2000; Veenhoven, 2013). On the basis of the distinctions between opportunities and outcomes, as well as the distinctions between outer and inner qualities of life, Veenhoven (2013) proposed quality of life as a multidimensional concept, which consists of four layers of meanings, namely, livability, life-ability, appreciation of life and utility of life. Livability refers to the living environment and conditions, whereas life-ability refers to individuals' capability to solve problems in life. In terms of life outcomes, appreciation of life is related to the self-perceived value for one's self, whereas the utility of life is associated with individuals' value to their environment (Veenhoven, 2013). Appreciation of life is associated with individuals' inner qualities of life and is therefore linked with psychological concepts, such as SWB, happiness and life satisfaction (Veenhoven, 2013).

SWB is taken from an individual's self-perceived view of quality of life, thus it falls into the concept of appreciation of life. According to Diener, Suh, Lucas, and Smith (1999), the conceptualization of SWB is as follows: 'subjective wellbeing is a broad category of phenomena that includes people's emotional responses, domain satisfactions and global judgments of life satisfaction' (Diener et al., 1999, p. 277). The domain and life satisfactions are more cognitive in nature and are often considered a trait, whereas emotional responses are more like a state. A good quality of life is represented by high life and domain satisfactions, high positive feelings and low negative feelings (Diener et al., 1999).

Research on travel and SWB

As stated previously, travel is posited to promote SWB (Chen & Petrick, 2013; Uysal et al., 2016). Several scholars have suggested that travel could significantly increase the SWB of individuals by providing opportunities to engage in memorable and pleasant experiences (de Bloom et al., 2011; Nawijn et al., 2010; Nawijn & Veenhoven, 2011). With regard to the underpinnings of the contribution of travel to SWB, the bottom-up spillover theory suggests that activities and experiences that are stored in concrete psychological domains shape individuals' contentment with specific life domains, and levels of contentment with various life domains interact simultaneously to form global life satisfaction (Kruger, 2012). The spillover effect suggests that satisfaction can be transferred from the most concrete domain to the most abstract life domain. Along with this process, contentment with specific life domains can be mediated by experiencing quality and activities associated with abstract life domains. In addition, emotions and feelings that accompany life events influence how individuals evaluate various life domains (Kruger, 2012). In the tourism literature, Neal et al. (2007) applied a theoretical framework guided by the bottom-up spillover theory to examine travel benefits and found that satisfactory travel experiences could increase travelers' contentment with specific life domains, as well as overall life satisfaction.

However, the benefits of travel on SWB may fluctuate at different stages across travel (de Bloom et al., 2010; Nawijn et al., 2010). In a comparison of travelers' SWB prior to and after vacations, researchers reported that participants felt happier after taking vacations (Dolnicar, Yanamandram, & Cliff, 2012; Gilbert & Abdullah, 2004; Nawijn et al., 2010; Pols & Kroon, 2007). However, the positive effects of travel do not last long. A fade-out stage where the benefits generated by travel disappear gradually has been observed (de Bloom et al., 2010; Kuhnel & Sonnentag, 2011; Nawijn, 2011b). The positive effects of traveling on the SWB of individuals could decline due to the workload assumed by travelers upon returning to work. The effects of traveling might last no more than one month (de Bloom et al., 2010; Kuhnel & Sonnentag, 2011; Pols & Kroon, 2007).

Influence of family travel on SWB

Zabriskie and McCormick (2003) categorized family leisure activities as core and balance activities. Core activities refer to 'common, everyday, low-cost, relatively accessible, and often home-based activities that many families do frequently' (Zabriskie & McCormick, 2003, p. 168). By contrast, balance activities are represented by 'less common, less frequent, more out of the ordinary, and usually not home-based activities thus providing novel experiences' (Zabriskie & McCormick, 2003, p. 168). Family travel fits in the balance activity category, and it addresses the element of unpredictability or novelty, which requires family members to negotiate and adapt to a new environment that is distinct from everyday life. Although family travel takes place away from the everyday environment, travel experiences are still within the borders of a family system (Zabriskie & McCormick, 2003). Thus, family members are in an interactive system in which they influence one another when adapting to a new environment (Fu et al., 2014).

One stream of studies has revealed that family travel positively influences well-being, especially family well-being (Durko & Petrick, 2013; Lehto et al., 2009; Shaw, Havitz, & Delamere, 2008). In particular, parents and children could optimize their relationships and enhance family cohesion during family travel (Lehto et al., 2009). Smith (1997) suggested that shared leisure activities during family travel create a unique experience that teaches children how to share and get along with others and develop loyalty to their family. Shaw et al. (2008) pointed out that family bonds are intensified through vacationing as vacations could create long-lasting memories that might generate meanings in the future and play a crucial role in future decision making (Shaw et al., 2008). In addition, Lehto et al. (2009) indicated that traveling with family members is perceived as quality time well spent by leisure travelers. In particular, family members could interact with one another through participation in various activities during family vacations. As a result, family ties are strengthened through the enhancement of connections and communications among family members (Lehto et al., 2009).

Although idealized family vacations are believed to be highly beneficial in improving family functioning and parent–child relationships, another stream of studies has addressed the potential negative outcomes of family travel on well-being (Rosenblatt & Russell, 1975; Shaw, 1997). Rosenblatt and Russell (1975) elaborated the fluidity of family dynamics when families are on the road and proposed that dealing with interpersonal problems is challenging due to the fact that family members have more shared territories and less personal space on the road than at home (Rosenblatt & Russell, 1975). In addition, the styles of child rearing and applications of family rules when on the road might not be the same as those at home, and such a difference could change the interactive dynamics between parents and children. Therefore, interactions within a family system might adversely influence family cohesiveness and increase individual stress. Moreover, recreational activities might not be created equally in facilitating or impeding the psychological and physical needs of individuals during family travel (Lehto et al., 2012). In other words, given that family members might have their particular personal pursuits during a trip, family leisure activities bring about challenges in establishing agreements with all family members. Ideally, family travel is thought to be an extended time spent with the family for relaxation and recovery, but combining the desires of different family members, such as parents and children and young and older children, sometimes turns out to be a challenge (Gram, 2005; Schanzel & Smith, 2014). Therefore, family travel may arguably cause conflicts and stress and therefore have a negative influence on the well-being of travelers.

Research on family travel and children's well-being

The social study on childhood addresses the need to reposition children as subjects rather than objects (Shaw, 1997). As a result, researchers dealing with family travel have realized that children are active agents in understanding family travel experiences (Havitz et al., 2010; Hilbrecht, Shaw, Delamere, & Havitz, 2008; Nickerson & Jurowski, 2001). However, scant attention has been paid to the potential benefits of family travel from children's perspectives in the tourism literature.

The need to include children's voices into family travel research is threefold. First, the influence of family travel on children might be different from that on parents due to family dynamics during family travel. In family leisure settings, children's activities are often chosen by their parents. In the absence of freedom to choose leisure activities, family leisure activities seem to be purposely arranged for children; thus, such activities could decrease children's sense of self-determination and influence their SWB in a negative way (Coleman & Iso-Ahola, 1993; Shaw & Dawson, 2001). As a result, children's participation in leisure might not always result in positive effects (Larson, Gillman, & Richards, 1997). Second, parents and children might seek significantly different benefits from family travel (Fu et al., 2014). In particular, parents, especially mothers, believe that family travel contributes to the family functioning in various ways, but children do not find that they benefit from family travel as much as their parents do. Several empirical studies have demonstrated that children's experiences and expectations in relation to family travel differ from those of their parents. Gram (2005) investigated the travel motives of children and parents and reported that children expect more fun and activities from a vacation, whereas parents appear to look for relaxation but still want to stay together with family members (Gram, 2005). Third, differences in the attitudes of parents and children toward the relationship between family leisure and family life satisfaction have been observed. Zabriskie and McCormick (2003) examined the influence of family leisure involvement on family satisfaction from the perspectives of parents, children, and the global family. The findings indicate that family leisure involvement could only significantly predict family satisfaction from the parents' perspective but not from the children's perspective. The conclusions suggest that previously proposed models in the tourism literature could not successfully explain children's experiences in family travel settings.

Given the reasons mentioned above, the influence of family travel on the SWB of children deserves close attention. This study could fill the research gaps in the following aspects. First, little research has been conducted to examine whether traveling during family holidays could increase the SWB of children. This study could provide insights by focusing on the perspectives of adolescent children and addressing the influence of travel on SWB to adolescents.

Second, extant research has pointed out the potential benefits of family travel on family well-being (Lehto et al., 2009; Shaw et al., 2008) and also argued the potential stress and conflicts that family travel may cause (Rosenblatt & Russell, 1975; Shaw, 1997). However, examinations of the influence of travel on the SWB of travelers in family travel settings remain few. The current research could provide empirical evidence of the influence of family travel on the SWB of travelers. Third, extant research on travel and SWB has been mostly conducted in Western contexts (de Bloom et al., 2011; Nawijn et al., 2010; Neal et al., 2007). Research has suggested that people from Eastern cultures perceived happiness differently compare with people from Western cultures (Diener, Suh, Smith, & Shao, 1995). Results gleaned from the current study could provide salient insights into how family holiday travel influences adolescents' SWB in an Eastern cultural context. The influence of family travel on the SWB of adolescent indicated in this paper may be specific to Chinese society, but many concerns are likely to be relevant to East Asian social and cultural contexts.

Methodology

Research design

By using Labor Day in China as an experimental context, a longitudinal research design was employed. To detect the changes in the self-reported well-being of adolescents, their SWB scores were measured prior to and after the family holiday. The current research sought to examine the influence of family holidays, especially family travel, on the SWB of Chinese adolescents. Hence, a comparison of the SWB of traveling and non-traveling adolescents was a major consideration in this study. Respondents who traveled were treated as the travel group, whereas those who did not travel were treated as the control group. Data collection was conducted at three stages. The first stage was initiated one week before Labor Day to gain a baseline of the SWB of the participants. The second stage was initiated the following week when the participants returned to school after Labor Day. The third stage took place one month after Labor Day to re-examine the SWB of the participants and test the remaining effects of the family holiday on their SWB.

To understand the influence of family holiday travel on the SWB of children (instead of their parents), Chinese middle school students aged 12–15 years were included as participants in this study. The reasons why middle school students were chosen as the targets are twofold. First, this study applied a structured survey method to collect data; hence, middle school students were old enough and sufficiently articulate to grasp abstract concepts, such as well-being. Previous research has provided support for the validity of the application of well-being measures to adolescents from grades 7 to 9 (Gilman et al., 2008; Huebner, 1991, 1994; Laurent et al., 1999). Second, the influence of family holiday travel on adolescents' SWB is of particular interest in this study. However, high school students might spend more time with their friends than with parents during holidays. Thus, this study chose to target middle school students as potential participants.

Data collection

The sample was drawn from two public middle schools in Qingdao, a large city located in the eastern part of Mainland China. With regard to participant recruitment, three classes from each grade were designated by each school (grades 7–9) to take the survey. All the students from the designated classes were treated as potential participants of this study. As a result, 675 students were recruited, that is, 360 students were from school A, and 315 students were from school B. The questionnaires were distributed during lunch breaks. As the study surveyed the participants at three stages, the participants were asked to indicate their student ID on each of the surveys so that respondent information could be paired up. The surveys for the first stage were distributed on 25 April 2016. Those for the second stage were distributed on 3 May 2016. Those for the third stage were distributed on 5 June 2016.

The questionnaire included the personal basic information of the respondents and the assessment of SWB. In particular, respondents' SWB was asked across the three stages of data collection, whereas the personal information of the participants was only sought at the first stage (i.e. sex, grade, if they have travel plans). In addition, the students were asked to confirm if they had traveled during Labor Day at the second stage. Diener et al. (1999) pointed out that SWB is a construct consisting of a cognitive evaluation of life and domain satisfactions and emotional components with the presence of a positive affect and absence of a negative affect. Therefore, this study measured SWB in three parts: global life satisfaction, contentment with specific life domains, and affect, which are the most often used SWB measures in the tourism literature (Chen, Lehto, & Cai, 2013; Gilbert & Abdullah, 2004; Nawijn et al., 2010).

The Students' Life Satisfaction Scale (SLSS; Huebner, 1991) was applied to examine the global life satisfaction of adolescents. The SLSS comprises seven items, such as 'My life is going well'. The item format requires participants to indicate how much they agree with statements regarding their SWB on a five-point Likert scale. In addition, the Multidimensional Students' Life Satisfaction Scale (MSLSS; Huebner, 1994) was used to measure the contentment of the participants with various life domains. In this study, the measures of the MSLSS were modified as follows. First, in the MSLSS (Huebner, 1994), 40 items were used, thereby making the MSLSS lengthy. Measuring contentment with specific life domains was not the only purpose of this study; hence, a short version of the MSLSS was applied. In Huebner's (1994) study, a scale of the MSLSS was developed on the basis of factor analysis. To make a short version of the MSLSS, this study selected three items of each domain with the highest factor loadings reflected in the results of the factor analysis in the study of Huebner (1994). Second, the MSLSS has five domains, namely, family, friends, school, living environment and self. The purpose of this study was to explore how family travel influences the SWB of adolescents; thus, the contentment of participants with their leisure life was added to the MSLSS as the sixth domain. Questions on the contentment of the participants with their leisure life were adopted from the study of Neal et al. (2007). Consequently, all 18 items, for example, 'I enjoy being at home with my family', were assessed on a five-point Likert scale, where 1 = strongly disagree and 5 = strongly agree. Moreover, this study used the Positive and Negative Affect Scale for Children (PANAS-C; Laurent et al., 1999) to assess the affect of adolescent students. The PANAS-C contains 12 positive affect items (e.g. interested and excited) and 15 negative affect items (e.g. sad and frightened) for evaluating affect. The current study instructed respondents to indicate how often they had felt those emotions recently on five-point Likert scales.

The original survey was developed in English on the basis of the literature review. Then, the researcher, who is bilingual in English (second language) and Chinese (first language), translated the survey from English to Chinese. To increase the accuracy of the translation, a bilingual research assistant back-translated the survey from Chinese into English. Finally, the researcher compared the language of the back-translation with the original version to check whether the contents were conveyed successfully between the two languages.

Data analyses

A variety of statistics were employed to analyze the data. First, descriptive analysis was applied to describe the characteristics of the sample according to reporting frequencies and percentages. Second, a series of Mixed and Repeated Measures of Analysis of Variance (ANOVA) were used to measure the interactions of time and travel on the SWB of the participants. Third, post-hoc tests were conducted to examine the simplified effects of time and travel on the SWB of the respondents. To ensure the assumptions of the statistics, Greenhouse–Geisser corrections were used to compensate for violations of assumptions of sphericity test results (Greenhouse & Geisser, 1959). In addition, Leneve's test was adopted to test the homogeneity of the variances (Brown & Forsythe, 1974).

Results

Profile of the data

A total of 675 students were recruited to participate in this study, and 606 of them returned their surveys across the three stages. However, the valid sample size was found to be 518 (76.7%) after excluding the respondents who turned in their survey responses with more than one section of missing values. The gender of the sample was equally split (Table 1). With regard to student grades, the numbers of valid surveys from grades 7 and 8 were twice that from grade 9. Overall, less than one quarter of the respondents (23.7%) traveled during Labor Day. Comparatively, more students from grade 7 traveled during Labor Day (36.8%), and only one in seven respondents from grades 8 and 9 traveled during Labor Day. Nevertheless, the gender differences between the traveling and non-traveling adolescents were not obvious.

Examinations of the time effect of family travel on the SWB of adolescents

Before testing the research questions, a series of mixed and repeated measures of ANOVA was used to examine if there are interaction effects in terms of time (three stages) and travel

Table 1. Profile of valid respondents.

		Frequency (%)		
		Non-travel	Travel	Total
Grade	7	127 (63.2)	74 (36.8)	201
	8	181 (84.2)	34 (15.8)	215
	9	87 (85.3)	15 (14.7)	102
Sex	Female	193 (73.4)	70 (26.6)	263
	Male	202 (79.2)	53 (20.8)	255
Total		395 (76.3)	123 (23.7)	518 (100)

on each construct (i.e. global life satisfaction; contentment with family, friends, school, living environment, self and leisure; and positive and negative affects). In particular, the value of endogenous variables were represented by the composite means of observed items. Specifically, the value for global life satisfaction was the composite mean of the seven items on the basis of the SLSS, the values for contentment with specific life domains were the composite means of the three observed items of each domain (i.e. family, friends, school, living environment, self, and leisure) of the MSLSS, and the value for positive affect was the composite mean of the 12 positive affect items and for negative affect was the composite mean of the 15 negative affect items on the basis of the PANAS-C.

Mauchly's test of sphericity was applied and revealed that all the models violated the assumptions of sphericity ($p < 0.001$). Thus, Greenhouse–Keiser's corrections were used to interpret within-group test results (Greenhouse & Geisser, 1959). Moreover, the results of Levene's tests of contentment with school life at the first and second stages were significant ($p < 0.05$). The error variances of the dependent variables were not equal across groups (Brown & Forsythe, 1974). The result could be affected by the differences in the samples from the travel and control groups. No significant interaction effects on satisfaction with school life were found. Thus, this current study did not address the interaction effects on the school life satisfaction of adolescents. In this sense, the misinterpretation of the results was reduced.

Table 2 displays the results of the interaction effects of time and travel on each construct (i.e. global life satisfaction; contentment with family, friends, school, living environment, self and leisure; and positive and negative affects). The results of the repeated measures of ANOVA indicated significant interaction effects in terms of time and travel on the SWB measures of global life satisfaction ($p < 0.01$), contentment with friends ($p < 0.05$) and positive ($p < 0.01$) and negative ($p < 0.001$) affects across the three stages. In particular, the partial eta squared for the interaction effects suggested small interaction effects of time and travel on global life satisfaction ($\eta = 0.017$), contentment with friendship ($\eta = 0.01$) and positive affect ($\eta = 0.016$). It also indicated the large interaction effect of time and travel on negative affect ($\eta = 0.371$).

To examine the time effects of family travel on the SWB of Chinese adolescent travelers, post-hoc tests (pairwise comparisons) were conducted to examine the simplified effects of time on the SWB of adolescent students. The marginal means of the SWB of the adolescents at the three stages were compared, whereas the travel group was controlled. The results are shown in Table 3. Estimated marginal means revealed that the global life

Table 2. Interaction effects of time and travel on the measures of SWB.

	Measure	df	F	Sig.	η^2
Time * Travel	Global life satisfaction	1.81	6.72	**0.002**	0.017
	Family life	1.90	0.42	0.650	0.001
	Friends	1.92	3.85	**0.023**	0.010
	School life	1.95	1.27	0.282	0.003
	Living environment	1.89	0.71	0.484	0.002
	Self	1.92	2.61	0.077	0.006
	Leisure life	1.93	2.85	0.060	0.007
	Positive affect	1.97	6.65	**0.001**	0.016
	Negative affect	1.78	236.02	**0.000**	0.371

Note. Results are based on the Greenhouse-Geisser corrections.
The bold indicates that the effects are significant.

Table 3. Pairwise comparisons of simplified effect of time on SWB.

Measure	Group	Stage 1	Stage 2	Stage 3
Global life satisfaction	Non-travel	3.94[a]	3.99[a]	4.03[a]
	Travel	**4.07[b]**	**4.35[a]**	**4.06[b]**
Family life	Non-travel	4.20[a]	4.12[a]	4.16[a]
	Travel	4.38[a]	4.34[a]	4.30[a]
Friends	Non-travel	**4.50[a]**	**4.35[b]**	**4.30[b]**
	Travel	4.58[a]	4.68[a]	4.54[a]
School life	Non-travel	3.75[a]	3.77[a]	3.86[a]
	Travel	**3.92[b]**	**4.11[a]**	**4.06[ab]**
Living environment	Non-travel	4.21[a]	4.19[a]	4.17[a]
	Travel	4.38[a]	4.45[a]	4.32[a]
Self	Non-travel	4.11[a]	4.14[a]	4.15[a]
	Travel	**4.19[b]**	**4.40[a]**	**4.23[b]**
Leisure life	Non-travel	**3.75[b]**	**3.84[ab]**	**3.86[a]**
	Travel	**4.05[b]**	**4.33[a]**	**4.13[b]**
Positive affect	Non-travel	4.01[a]	4.02[a]	4.08[a]
	Travel	**4.20[b]**	**4.47[a]**	**4.29[b]**
Negative affect	Non-travel	**2.28[b]**	**4.03[a]**	**2.11[c]**
	Travel	**3.78[a]**	**1.83[b]**	**1.86[b]**

Note. Stage 1 = Before the Labor Holiday; Stage 2 = Right after the Labor Holiday; and Stage 3 = One month after the Labor Holiday.
a, b, c represent the results of post-hoc tests of pairwise comparisons between every two stages, where a > b > c.

satisfaction, contentment with school life, self, leisure life and positive affect right after the holiday of adolescent travelers were significantly higher than those before the holiday and one month after the holiday. As for the negative affect of adolescents, it was significantly low right after the holiday. These findings suggested that the SWB of the adolescents who traveled during Labor Day significantly increased, but the benefits on well-being associated with family travel were not sustained. For the adolescents who did not travel during Labor Day, their contentment with friendship was significantly low right after the holiday. In addition, the negative affect of non-traveling adolescents significantly increased after the holiday. Thus, the family holiday could not significantly improve the SWB of the adolescent students who did not travel during the holiday, but it did decrease the contentment of non-traveling adolescents with friends and increase their negative affect.

With regard to research question 1, this study found a short-term lift-up effect of family holiday travel on the SWB of adolescents. This finding was only applied to the adolescents who traveled during the family holiday. With respect to research question 2, the SWB measures for the travel group one month after the holiday significantly dropped to the pre-travel stage. Consequently, this study could not support a long-term effect of family travel on the SWB of adolescent students. In other words, the well-being-related benefits that the adolescent students gained during the family travel were not sustained.

Examinations of the influence of travel on the SWB of adolescents

To address whether SWB between adolescents who traveled and those adolescents who did not travel during Labor Day had any differences (research question 3), this study conducted post-hoc tests and examined the simplified effect of travel as an independent factor on the SWB of the adolescents. The marginal means of SWB between traveling and non-traveling adolescents were compared, with time being controlled. The mean difference suggested a significant difference ($p < 0.05$, Table 4) among all SWB measures of the traveling and non-

Table 4. Pairwise comparisons of simplified effect of travel on SWB.

Measure	Stage	Travel	Non-travel	Mean difference
Global life satisfaction	1	4.07[a]	3.94[a]	0.13
	2	**4.35[a]**	**3.99[b]**	**0.36***
	3	4.06[a]	4.03[a]	0.04
Family life	1	4.38[a]	4.20[a]	0.18
	2	**4.34[a]**	**4.12[b]**	**0.23***
	3	4.30[a]	4.16[a]	0.14
Friends	1	4.58[a]	4.50[a]	0.08
	2	**4.68[a]**	**4.35[b]**	**0.33***
	3	**4.66[a]**	**4.30[b]**	**0.24***
School life	1	3.92[a]	3.75[a]	0.17
	2	**4.11[a]**	**3.77[b]**	**0.34***
	3	4.06[a]	3.86[a]	0.21
Living environment	1	4.38[a]	4.21[a]	0.17
	2	**4.45[a]**	**4.19[b]**	**0.25***
	3	4.32[a]	4.17[a]	0.15
Self	1	4.19[a]	4.11[a]	0.07
	2	**4.40[a]**	**4.14[b]**	**0.26***
	3	4.23[a]	4.15[a]	0.08
Leisure life	1	**4.05[a]**	**3.75[b]**	**0.30***
	2	**4.33[a]**	**3.84[b]**	**0.50***
	3	**4.13[a]**	**3.86[b]**	**0.27***
Positive affect	1	**4.20[a]**	**4.01[b]**	**0.19***
	2	**4.47[a]**	**4.02[b]**	**0.46***
	3	**4.29[a]**	**4.08[b]**	**0.21***
Negative affect	1	**3.78[a]**	**2.28[b]**	**1.50***
	2	**1.83[b]**	**4.03[a]**	**−2.20***
	3	**1.86[b]**	**2.11[a]**	**−0.25***

Note. Stage 1 = Before the Labor Holiday; Stage 2 = Right after the Labor Holiday; and Stage 3 = One month after the Labor Holiday.
a, b, c represent the results of post-hoc tests of pairwise comparisons between every two stages, where a > b > c.

traveling adolescents when the students returned to school (second stage). In addition, a significant difference was found between contentment with leisure life and positive and negative affects of the traveling and non-traveling adolescents across the three stages. Thus, the results suggested that the adolescents who traveled during the family holiday had significantly higher life satisfaction, higher contentment with various life domains, and greater positive emotions and they experienced significantly lower negative feelings compared with the adolescent students who did not travel. With regard to research question 3, travel could be a potential factor enhancing life and domain satisfactions of adolescents and adjusting their emotional well-being effectively.

Discussions

This study identified a significant interaction effect of time and travel during family holiday on the SWB of Chinese adolescents. This effect is particularly reflected on the global life satisfaction, contentment with friends, and positive and negative affects of adolescent students. The current study further examined the simplified effect of time and travel on the SWB of adolescents, and suggests that family travel during holidays may potentially influence the SWB of adolescent students. Specifically, a short-term lift-up effect of family travel on the SWB of Chinese adolescent travelers was found. In particular, the adolescents who traveled during the family holiday experienced an increase in global life satisfaction; contentment with school, self and leisure life; and positive affect, and also experienced a decreased in negative affect after the holiday. This study echoes to extant studies discussing the links between family leisure and well-being of adolescents. First, through family leisure participation, children are able to develop the identity of themselves, their families and cultures in a supportive environment, which is considered to be helpful to the personal development and well-being maintenance of children (Caldwell & Darling, 1999; Iso-Ahola & Crowley, 1991; Kleiber & Kirshnit, 1991), and thus, adolescent students may develop a high contentment with self. In addition, family travel provides chances to stay away from daily routines, enabling family members to engage in pleasurable diversionary activities. As a result, adolescents can take a break from their busy school life, which may reduce intense study pressure, enhance their satisfaction with school and leisure lives, and induce positive affect and reduce negative affect.

On the contrary, the long-term effect of family travel on the SWB of adolescent travelers is not supported by this study. In other words, the benefits of family travel are not sustainable; instead, the SWB of adolescent travelers drops to pre-travel level one month after the holiday. Previous studies have demonstrated that the benefits of travel on the SWB of working adults are decreased by their workload when they return to work (Chen et al., 2013; de Bloom et al., 2010, 2011; Kuhnel & Sonnentag, 2011). This study provides a consistent finding that Chinese adolescents who travel experience a significant decrease in the measures of global life satisfaction, contentment with school life, self, leisure life and positive affect and an increase in negative affect one month after the holiday. However, Chinese middle school students experience intense pressure immediately when they return to school. On the basis of free conversations between the researcher and the student participants, the respondents mentioned having 10-hour classes from Monday to Friday. Moreover, they need to spend at least three hours to finish homework after school. More important, middle school students have quizzes almost every day. Thus, the dynamics of the fluctuations of the SWB of Chinese adolescents may drop immediately rather than gradually after holidays. Future researchers can measure the SWB of adolescents several times to trace changes in their SWB after holidays. Furthermore, given the 'fade-out' effects of benefits generated from family travel, discovering how to sustain benefits for long periods of time after holidays is important. Future studies can explore the factors that contribute to the maintenance of the beneficial effects of family travel on the SWB of adolescents.

In the current study, the pairwise comparisons suggested significant differences between the SWB of the traveling and non-traveling adolescents upon their return to school. No significant differences were found for the pre-holiday and post-holiday stages. This finding could suggest that family travel during holidays can significantly influence the SWB of adolescent students. However, other factors can simultaneously affect the SWB of adolescents. For instance, many factors influence family travel decisions, such as family disposable money and available time for both parents and children. Thus, recognizing the benefits of travel on the promotion of individual well-being is helpful, but valuing other factors is equally important.

Furthermore, this study found that two-thirds of Chinese adolescents do not take trips during holidays. Most Chinese adolescents decide to spend holidays to study, either reviewing independently or attending tutorials at private institutions. Two possible reasons can explain the low rate of travel during the family holiday. First, the length of the holiday is too short for extended travel, and so families choose to relax at home rather than take trips. Second, Labor Day is celebrated one month before the final examination of middle school. Thus, most students prefer to spend their holiday on course reviews rather than on traveling with families. Future research can study the holiday experiences of adolescents at an earlier time of the academic year, such as the National and Spring Festival Holidays, which may provide different perspectives. Existing studies have mainly focused on describing the changes in the SWB of travel groups (Chen et al., 2013; Gilbert & Abdullah, 2004; Nawijn, 2011b). Meanwhile, examinations of SWB fluctuations for non-traveling groups are hardly found. Indeed, adolescents who do not travel during family holidays have largely been underexplored. Non-traveling adolescents do not experience the benefits of family holidays on their SWB. Thus, their holiday experiences should be understood, and the ways that can promote their quality of life should be explored.

Lastly, this study found that family satisfaction across the three stages does not significantly differ between traveling and non-traveling adolescents. Adolescent students may spend extended time with their parents during family holidays. Previous studies have proposed that a lot of potential opportunities can encourage intra-family communications and interactions to increase family bonds during family holidays (Lehto et al., 2009). However, the current study cannot support the notion that family holidays effectively increase the family satisfaction of adolescents. This study found that contentment with family life of Chinese adolescents is below the average level across the three stages. One potential reason may be that adolescent students do not fully take advantage of their time with families during holidays. Future studies can be developed to closely study the dynamics of family interactions between parents and children during family holidays and explore the relationships of family interactions and SWB of adolescents in the context of family holidays.

Conclusion

This study examined the influence of family holiday travel on the SWB of Chinese adolescents. Overall, two-thirds of the respondents did not travel during the family holiday. For the Chinese adolescents who traveled over Labor Day, SWB significantly increased, particularly in terms of overall life satisfaction, contentment with school, self and leisure life, and emotional well-being. The SWB of the non-traveling adolescents did not change significantly across the holiday. In addition, the SWB of the adolescents who traveled during the holiday was significantly higher than that of the adolescents who did not travel when they returned to school. Finally, the results suggested that travel has no long-term effect on the SWB of adolescents. In other words, the potential role that the family travel played in increasing the SWB of the students diminished after the holiday.

The findings of this study have theoretical implications. The current study advances our knowledge of the influence of family travel on the SWB of adolescents. Family travel can potentially play a beneficial role in enhancing the global life satisfaction, contentment with leisure life, school life, self and emotional well-being of adolescents. This study fills in important research gaps through demonstrations involving the adolescent group, which add value and make our understanding comprehensive. Moreover, few studies have empirically examined the influence of travel on the SWB of travelers in family travel settings. This study suggested that family travel may arguably play a beneficial role to enhance travelers' global life satisfaction and emotional well-being. However, the beneficial influence of family travel on the contentment of adolescents with family life has not been observed. This study suggests that the influence of travel on SWB in family travel settings may not be consistent with the results of other types of travel. Thus, future research should consider the context of family travel when discussing the influence of travel on travelers' SWB. In terms of practical implications, the pressure to study can be one of the main factors that discourage Chinese adolescents from traveling with their families and influence their SWB in a negative way. This study suggests that schools and parents give adolescent students a break during holidays by removing pressure and taking advantage of holiday time to arrange trips. Moreover, when adolescent students return to school after holidays, their participation in leisure activities should be encouraged to sustain the beneficial effects of family travel on their SWB. This study addressed the influence of family travel on the SWB of adolescents and suggested that travel can potentially play a beneficial role in enhancing the SWB of adolescent students. Parents, schools, and the government should be aware that the well-being and quality of life of adolescent students deserve as much attention as that received by their academic achievements.

Disclosure statement

No potential conflict of interest was reported by the authors.

References

Brown, M. B., & Forsythe, A. B. (1974). Robust tests for the equality of variances. *Journal of the American Statistical Association*, *69*(346), 364–367.

Caldwell, L. L., & Darling, N. (1999). Leisure context, parental control, and resistance to peer pressure as predictors of adolescent partying and substance use: An ecological perspective. *Journal of Leisure Research*, *31*(1), 57–77.

Carr, N. (2006). A comparison of adolescents' and parents' holiday motivations and desires. *Tourism and Hospitality Research*, *6*(2), 129–142.

Chen, C. C., & Petrick, J. F. (2013). Health and wellness benefits of travel experiences: A literature review. *Journal of Travel Research, 52*(6), 709–719.

Chen, Y., Lehto, X. Y., & Cai, L. (2013). Vacation and well-being: A study of Chinese tourists. *Annals of Tourism Research, 42*, 284–310.

Coleman, D., & Iso-Ahola, S. E. (1993). Leisure and health: The role of social support and self-determination. *Journal of Leisure Research, 25*(2), 111–129.

de Bloom, J., Geurts, S. A., Sonnentag, S., Taris, T., de Weerth, C., & Kompier, M. A. (2011). How does a vacation from work affect employee health and well-being? *Psychology & Health, 26*(12), 1606–1622.

de Bloom, J., Geurts, S. A., Taris, T. W., Sonnentag, S., de Weerth, C., & Kompier, M. A. (2010). Effects of vacation from work on health and well-being: Lots of fun, quickly gone. *Work & Stress, 24*(2), 196–216.

Diener, E., Suh, E., Lucas, R. E., & Smith, H. L. (1999). Subjective well-being: Three decades of progress. *Psychological Bulletin, 125*(2), 276–302.

Diener, E., & Suh, E. M. (Ed.). (2000). *Culture and subjective well-being*. Cambridge, MA: The MIT Press.

Diener, E., Suh, E. M., Smith, H., & Shao, L. (1995). National differences in reported subjective well-being: Why do they occur? *Social Indicators Research, 34*(1), 7–32.

Dolnicar, S., Yanamandram, V., & Cliff, K. (2012). The contribution of vacations to quality of life. *Annals of Tourism Research, 39*(1), 59–83.

Durko, A. M., & Petrick, J. F. (2013). Family and relationship benefits of travel experiences: A literature review. *Journal of Travel Research, 52*(6), 720–730.

Fritz, C., & Sonnentag, S. (2006). Recovery, well-being, and performance-related outcomes: The role of workload and vacation experiences. *Journal of Applied Psychology, 91*(4), 936–945.

Fu, X., Lehto, X., & Park, O. (2014). What does vacation do to our family? Contrasting the perspectives of parents and children. *Journal of Travel & Tourism Marketing, 31*(4), 461–475.

Gilbert, D., & Abdullah, J. (2004). Holiday-taking and the sense of well-being. *Annals of Tourism Research, 31*(1), 103–121.

Gilman, R., Huebner, E. S., Tian, L., ark, N., O'Byrne, J., Schiff, M., … Langknecht, H. (2008). Cross-national adolescent multidimensional life satisfaction reports: Analyses of mean scores and response style differences. *Journal of Youth and Adolescence, 37*(2), 142–154.

Gram, M. (2005). Family holidays: A qualitative analysis of family holiday experiences. *Scandinavian Journal of Hospitality and Tourism, 5*(1), 2–22.

Greenhouse, S. W., & Geisser, S. (1959). On methods in the analysis of profile data. *Psychometrika, 24*(2), 95–112.

Havitz, M. E., Shaw, S. M., & Delamere, F. (2010). Marketing family vacations: What recreation professionals should know. *Parks & Recreation, 45*(6), 29–31.

Hilbrecht, M., Shaw, S. M., Delamere, F. M., & Havitz, M. E. (2008). Experiences, perspectives, and meanings of family vacations for children. *Leisure/Loisir, 32*(2), 541–571.

Hu, F., Ma, Y. H., Hu, L. M., Deng, X. L., & Mei, J. F. (2010). Chuzhongsheng zhuguan xingfugan yu jiating gongneng de guanxi [Study on the relationship of family functioning and subjective well-being of junior-high school students]. *Beijing Da Xue Xue Bao (Yi Xue Ban), 42*(3), 323–329.

Huebner, E. S. (1991). Initial development of the student's life satisfaction scale. *School Psychology International, 12*(3), 231–240.

Huebner, E. S. (1994). Preliminary development and validation of a multidimensional life satisfaction scale for children. *Psychological Assessment, 6*(2), 149–158.

Iso-Ahola, S. E., & Crowley, E. D. (1991). Adolescent substance abuse and leisure boredom. *Journal of Leisure Research, 23*(3), 260–271.

Kleiber, D. A., & Kirshnit, C. E. (1991). Sport involvement and identity formation. In L. Diamant (Ed.), *Mind-body maturity: Psychological approaches to sports, exercise, and fitness* (pp. 193–211). Washington, DC: Hemispudhererp.

Kruger, P. S. (2012). Perceptions of tourism impacts and satisfaction with particular life domains. In M. Uysal, R. Perdue, & M. J. Sirgy (Eds.), *Handbook of tourism and quality-of-life research:*

Enhancing the lives of tourists and residents of host communities (pp. 279–292). Dordrecht, Netherlands: Springer.
Kuhnel, J., & Sonnentag, S. (2011). How long do you benefit from vacation? A closer look at the fade-out of vacation effects. *Journal of Organizational Behavior, 32*(1), 125–143.
Larson, R. W., Gillman, S. A., & Richards, M. H. (1997). Divergent experiences of family leisure: Fathers, mothers, and young adolescents. *Journal of Leisure Research, 29*(1), 78–97.
Laurent, J., Catanzaro, S. J., Joiner Jr., T. E., Rudolph, K. D., Potter, K. I., Lambert, S., & Gathright, T. (1999). A measure of positive and negative affect for children: Scale development and preliminary validation. *Psychological Assessment, 11*(3), 326–338.
Lehto, X. Y., Choi, S., Lin, Y. C., & MacDermid, S. M. (2009). Vacation and family functioning. *Annals of Tourism Research, 36*(3), 459–479.
Lehto, X. Y., Lin, Y. C., Chen, Y., & Choi, S. (2012). Family vacation activities and family cohesion. *Journal of Travel & Tourism Marketing, 29*(8), 835–850.
Nawijn, J. (2011a). Determinants of daily happiness on vacation. *Journal of Travel Research, 50* (5), 559–566.
Nawijn, J. (2011b). Happiness through vacationing: Just a temporary boost or long-term benefits? *Journal of Happiness Studies, 12*(4), 651–665.
Nawijn, J., Marchand, M. A., Veenhoven, R., & Vingerhoets, A. J. (2010). Vacationers happier, but most not happier after a holiday. *Applied Research in Quality of Life, 5*(1), 35–47.
Nawijn, J., & Veenhoven, R. (2011). The effect of leisure activities on life satisfaction: The importance of holiday trips. In I. Brdar (Ed.), *The human pursuit of well-being* (pp. 39–53). Dordrecht, Netherlands: Springer.
Neal, J. D., Uysal, M., & Sirgy, M. J. (2007). The effect of tourism services on travelers' quality of life. *Journal of Travel Research, 46*(2), 154–163.
Nickerson, N. P., & Jurowski, C. (2001). The influence of children on vacation travel patterns. *Journal of Vacation Marketing, 7*(1), 19–30.
Pols, J., & Kroon, H. (2007). The importance of holiday trips for people with chronic mental health problems. *Psychiatric Services, 58*(2), 262–265.
Rosenblatt, P. C., & Russell, M. G. (1975). The social psychology of potential problems in family vacation travel. *The Family Coordinator, 24*, 209–215.
Schanzel, H. A., & Smith, K. A. (2014). The socialization of families away from home: Group dynamics and family functioning on holiday. *Leisure Sciences, 36*(2), 126–143.
Seymour, J., & McNamee, S. (2012). Being parented: Children and young people's engagement with parenting activities. In J. Walden & I. M. Kaminski (Eds.), *Learning from the children: Childhood, culture and identity in a changing world* (pp. 92–107). Oxford, England: Berghahn.
Shaw, S. M. (1997). Controversies and contradictions in family leisure: An analysis of conflicting paradigms. *Journal of Leisure Research, 29*(1), 98–112.
Shaw, S. M., & Dawson, D. (2001). Purposive leisure: Examining parental discourses on family activities. *Leisure Sciences, 23*(4), 217–231.
Shaw, S. M., Havitz, M. E., & Delamere, F. M. (2008). "I decided to invest in my kids' memories": Family vacations, memories, and the social construction of the family. *Tourism Culture & Communication, 8*(1), 13–26.
Sirgy, M. J., Kruger, P. S., Lee, D. J., & Grace, B. Y. (2011). How does a travel trip affect tourists' life satisfaction? *Journal of Travel Research, 50*(3), 261–275.
Small, J. (2008). The absence of childhood in tourism studies. *Annals of Tourism Research, 35*(3), 772–789.
Smith, D. S. (1997). Strengthening family values in the twenty-first century- home-centered recreation. *Journal of Physical Education, Recreation & Dance, 68*(8), 39–41.
Tian, L., Liu, B., Huang, S., & Huebner, E. S. (2013). Perceived social support and school well-being among Chinese early and middle adolescents: The mediational role of self-esteem. *Social Indicators Research, 113*(3), 991–1008.
Uysal, M., Sirgy, M. J., Woo, E., & Kim, H. L. (2016). Quality of life (QOL) and well-being research in tourism. *Tourism Management, 53*, 244–261.

Veenhoven, R. (2013). The four qualities of life ordering concepts and measures of the good life. In A. Delle Fave (Ed.), *The exploration of happiness* (pp. 195–226). Dordrecht, Netherlands: Springer.

Zabriskie, R. B., & McCormick, B. P. (2003). Parent and child perspectives of family leisure involvement and satisfaction with family life. *Journal of Leisure Research*, *35*(2), 163–189.

The Constructions of Family Holidays in Young Middle-class Malaysian Chinese Children

Catheryn Khoo-Lattimore and Mona Jihyun Yang

ABSTRACT
The absence of children in tourism research has been identified by past scholars but the rising attention on Asians consumers in tourism has also implied that the voices of Chinese children need to be heard. This study triangulates 14 young Malaysian Chinese children's drawings of their favorite family holidays with open-ended interview questions. The drawings were analyzed using the five thematic levels according to a model of analysis of children's drawings. Findings revealed the collectivistic Confucian value of family and education during family vacations, but also highlight what Chinese children conceptualized as fun holiday activities. A number of suggestions are provided for future research and for tourism practitioners to provide memorable holiday experiences for their young travelers.

马来西亚中产华裔儿童家庭假期的构建

摘要
学者指出过往的旅游研究中缺少儿童，亚洲旅游消费者对儿童的关注越多，亦意味着需要倾听华裔儿童的声音。本研究对十四名马来西亚华裔儿童的家庭假期画像进行了三角测量，并提供开放式访问问题。 根据儿童绘画分析模型，使用五个专题水平分析画像。 研究结果揭示了在休假期间集体主义的儒家家庭和教育价值观，同时也突出了华裔儿童将其概念化为有趣的假日活动。本文亦为未来的研究和旅游从业者提供建议。

Introduction

Tourism scholars have already begun to highlight the differences between Asian and Western travelers. Asian guests in hotels, for example, are not as familiar as non-Asians with complaint channels and are therefore less likely to complain on holidays (Ngai, Heung, Wong, & Chan, 2007). When compared with Americans, Asians are more forgiving of hotel service failures (Ekiz & Au, 2011) but expect a higher standard of service quality than their Western counterparts (Reisinger & Turner, 2002). Further, Reisinger and Turner (2002) found significant differences between Australian and

Asians in the tourism industry. More specifically, they pointed out that Asians are 'very much family oriented ... each member of the family depends on the other for security and protection' and that these family ties do not exist amongst Australians (Reisinger and Turner, 2002, p. 305). Evidence for this distinction in holiday behaviors between Asians and Westerners is growing in cross-cultural tourism studies. It has been found that learning, education and the acquisition of knowledge are especially valued by Asian adults during vacations (Ap, 2004; Wong., 1996) and are impressed upon their children during family holidays (Khoo-Lattimore, Prayag, & Cheah, 2015). Empirically, Asian parents have articulated demand for educational-related programs that range from swimming classes and horse riding to cooking and nature-related programs for their children at holiday sites, including accommodation (Khoo-Lattimore et al., 2015).

Given that research on family tourism is limited (Carr, 2011; Poria & Timothy, 2014; Schänzel, Yeoman, & Backer, 2012), and studies on Asian families more so (Kim, Choi, Agrusa, Wang, & Kim, 2010; Wang, Hsieh, Yeh, & Tsai, 2004), our understanding about Asian children on holidays is far from advanced. This study therefore aims to fill this gap by exploring the holiday experiences of young Malaysian Chinese children, by listening to their voices through a child-led research method. By doing so, this paper makes a contribution to the family tourism literature by identifying variables and constructs from some of the youngest members of families. By focusing on Chinese families, the findings will contribute to our theoretical understanding on meanings of holidays for Asian families. The rest of this paper presents a critique of current tourism and hospitality research on children. In doing so, it also highlights the potential influences of the Confucian value of 'family' (inherent in many Chinese communities) on how Chinese people experience family holidays. The methodology for the study is then explained, before the key findings are presented and discussed.

Literature review

Children in tourism

Existing studies often fail to recognize that young children are social actors that influence the vacation experience of their families (Schänzel et al., 2012). This could be due in part to the assumption that older children hold a stronger influence in their family holiday decision-making and younger children exert limited, if any, authority. A literature review of tourism studies involving children revealed that data for family tourism research has been mostly collected from parents (see for examples, Chen, Lehto, Behnke, & Tang, 2015; Howard & Madrigal, 1990; Khoo-Lattimore et al., 2015; Lehto, Choi, Lin, & MacDermid, 2009; Lugosi, Robinson, Golubovskaya, & Foley, 2016; Segumpan, Zahari, & Jamaluddin, 2008; Thornton, Shaw, & Williams, 1997; Turley, 2001) or adult caregivers (Gaines, Hubbard, Witte, & O'Neill, 2004). The omission of children from the samples is problematic in that it limits our knowledge of how children experience holidays, what they need and if they would like to holiday differently from their parents. As the dearth of children's voices has been highlighted, scholars have started to involve children in family tourism research. Table 1 summarizes existing studies that have done this and it highlights three significant observations.

Table 1. Summary of studies on children in tourism.

No	Study	Context	Children's Age	Method	Primary Findings
1	Canosa, Wilson, & Graham (2017)	Australians	Between 10 & 24	Filmmaking by 14 young people, some of whom were children	The viability of participatory visual method to engage young people as research respondents.
2	Blichfeldt et al. (2011)	Danes	Between 8 & 12	Drawing and focus group interviews with 89 children	Children do not passively submit to whatever choices their parents make
3	Carr (2006)	Australians	Between 12 & 17	54 parents and 66 children	Parents and children expressed different notions of ideal holidays
4	Cullingford (1995)	British	Between 7 & 11	Qualitative interviews with 160 children	Children view holidays as packages which include beach and entertainment.
5	Gram (2007)	Germans & Danish	Between 0 & 11	Qualitative interviews with 49 children	Parents and children have different views of the 'right matrix' of rest and activity.
6	Hilbrecht et al. (2008)	North Americans & Canadians	Between 5 & 16	Qualitative interviews with 24 children with their families	Children have different expectation of family vacations compared to their parents
7	Johns & Gyimóthy (2003)	Mostly Danes	No information was provided	Qualitative interviews with 35 families that included children	Many children mentioned 'fun', but would not say exactly what this meant for them.
8	Liang (2013)	Taiwanese	Between 12 & 18	Quantitative survey on children and their parents	Children's internet involvement and expertise has led to their having greater influences on their parents for tourism products purchasing via the Internet channel.
9	Nickerson and Jurowski (2001)	Americans	Between 10 & 17	Quantitative survey on adults and children	Children on vacation have opinions about their visit and these opinions may differ from those of their parents.
10	Schänzel (2012)	White New Zealanders	Between 6 & 16	Qualitative interviews with 20 parents and 20 children	Conflicts between children and parents on the definition of fun during holidays
11	Small (2008)	White Australians	12	Qualitative interviews with X girls	12 year olds enjoy resorts, and are focused on 'me' and 'my holiday'.
12	Tagg & Seaton (1995)	Belgian, French, British & Italian	Between 12 & 18	Quantitative surveys on parents and children	Children play a part on decision-making.
13	Therkelsen (2010)	Germans & Danes	Between 0 & 11	Qualitative interviews with parents and children	The influence of children is significant and develops from indirect to direct as the age of the children increases.

First, the majority of the children involved are older, typically of school age and commonly adolescents. Gram (2007) and Therkelsen (2010) were working off the same dataset (Gram & Therkelsen, 2003) and although children as young as 0 were involved, 'interviews were conducted without any significant input from children' (Therkelsen, 2010, p. 771). It can be derived from the literature that the youngest children interviewed were aged five (Hilbrecht, Shaw, Delamere, & Havitz, 2008) and six (Schänzel, 2012) years old. These scholars who have worked with younger children have reported problems in their attempts to extract quality data (Hilbrecht et al., 2008; Schänzel & Smith, 2014). Khoo-Lattimore (2015) attributes these difficulties and the prevalent absence of young children's voices to methodological challenges and outlined five considerations when researching young children.

A second observation is the persistent theme that children have their own opinions and ideals about what constitutes a good holiday, which differ from those of their parents. Carr (2006) highlights this contrast with a 13-year-old female respondent and her parent in his study – while the teenager described her perfect holiday as staying in a Hollywood motel with her friends, going to the beach, shopping, visiting theme parks and visiting celebrities, her mother described their ideal holiday as having a huge church gathering of friends and family in Broomshead, Australia with the children undertaking lots of beach activities. Dissimilarities such as this between children and parents were also emphasized by six other studies (Blichfeldt, Pedersen, Johansen, & Hansen, 2011; Gram, 2007; Hilbrecht et al., 2008; Nickerson & Jurowski, 2001; Schänzel, 2012) but given that they focus on older children, insights into younger children's holiday ideals are missing.

Finally, and more surprisingly, only one piece of research was conducted on Asian children (Liang, 2013). Liang (2013) found that Taiwanese teenagers have a significant influence on their parents in the context of purchasing tourism products online. Despite the growing research on Asia (Mura & Pahlevan Sharif, 2015), and on China, we still do not have enough data on what Chinese children want from their holidays, what family vacations mean for them and how they would like to experience leisure travel. Given that travel attitudes and behaviors developed in the early years have impacts on travel consumption behaviors in later life (Small, 2008; Todd Weaver, Moschis, & Davis, 2011), and coupled with the growing number of Chinese travelers as one of the largest tourism markets for the world (UNWTO, 2016), it is critical that tourism marketing scholars identify and/or strengthen new and existing frameworks and theories for understanding Chinese children.

Chinese children in family tourism

Although it has been established that the Chinese market is heterogeneous (Khoo-Lattimore, Yang, & Lai, 2016; Weaver et al., 2015), Confucian values characterize and explain behaviors observed in many Chinese societies (Yang, Khoo-Lattimore, & Arcodia, 2016), including but not limited to those in Mainland China, Taiwan, Japan, Korea, Hong Kong, Singapore, Indonesia and Malaysia. The influence of Confucianism on Chinese tourist behavior has been confirmed by many studies (Fu, Cai, & Lehto, 2015; Kwek & Lee, 2010; Mok & DeFranco, 2000; Tse & Hobson, 2008) but these have not included young Chinese tourists per se. For scholars researching family tourism, attention to this exclusion would be worthwhile because the family, in Confucianism, is the prototype of all social organizations (Mok & DeFranco, 2000). This means that a person is not primarily an individual but rather a member of a family, and so children are taught to suppress their individuality and to conform to the collectivistic goals of the family to maintain harmony.

Confucian values were alluded to in Khoo-Lattimore et al.'s (2015) study on Asian family tourism – their findings revealed that Malaysian Chinese parents discussed their preferences for educational activities at the holiday destination and accommodation – but the study collected data only from the mothers and not children. Therefore, although the literature acknowledges that Confucian values teach children to be polite to their seniors, respect their teachers and tolerate injustice (Hsieh & Scammon, 2015), the impacts of these values on how young Chinese children respond to family holidays are not yet known. We also know that subordination and harmony through hierarchical arrangements are key principals within Confucian values (Hsieh & Scammon, 2015) and that Chinese children influenced by Confucian values, 'tend to be more respective and receptive to their parents' arrangements in their daily lives in order to maintain harmony and the hierarchical relationship' (Lee, Wong, & Brown, 2015, p. 214). However, our knowledge of how and whether these values translate into travel behavior remains absent. We also recognize that Asian parents emphasize the value of education even during leisure and when on holidays, but whether or not their children conform as prescribed by their Confucian teachings or have different opinions, as Western children do, warrants further investigation. The aim of this study therefore is to address these gaps regarding Chinese children in family tourism research and asks the questions of how young Chinese children experience family holidays, and how they define fun. Underpinned by Confucianism, this study conceptually explores the reconstruction of family tourism through young children (whose voices remain marginal in tourism), and addresses the increasing call to shift discussions away from Eurocentric theories of travel consumption.

Methodology

Sampling

In order to address the gap on younger-aged children in existing family tourism research, this study will attempt to collect data from as young a child as possible. The literature provides some guidelines, and recommends five and six year olds as the most suitable youngest candidates for interviews or surveys because they are able to work on a task for more than an hour (Ruff & Lawson, 1990), are capable of recalling autobiographical memories (Fivush & Hamond, 1990), and can express their thoughts on complex issues such as family conflicts (Gross & Hayne, 1998). As such, the respondents for this study were sought from a private primary school in Malaysia. More specifically, students from its preschool classes were selected for this study because their ages (five and six) match those recommended by the literature. The help of class teachers was sought to hand-pick those who are more proficient in their language skills, and consent was sought from the students' parents through the school. Eighteen students were nominated, but only the 14 that were Chinese were included in this study. Although school fees for private schools are comparatively more expensive than public schools in Malaysia, not all parents of the children who attend them are necessarily wealthy. Many middle-class families, particularly Chinese middle-class families, are increasingly opting for private education for its perceived quality, comprehensiveness and international appeal (Malaysian Parents Choose International Schools Over National Schools These Days – What Are The Advantages?, 2015). A few private schools in Malaysia do charge exorbitant fees for the elite but the majority of them compete for the larger segment and offer competitive fees for middle-income families. The school selected for this study is an example of the latter.

Data collection method

Given that the aim and nature of the study are exploratory, a qualitative approach was adopted.

Data collection was based on a visual research method. The children were asked to draw a picture of their favorite holiday. Drawing is thought to be most suitable for preschool children as it is an activity that they are most familiar with, and typically enjoy. Barlow, Jolley, and Hallam (2011, p. 480) argued that because children's spontaneous and self-directed drawings reflect what they are most interested in, drawings are likely to elicit answers as to the most salient episodes associated with the topic of the drawing. Mitchell (2006) posited that drawings are a tool for engaging children to identify issues that are of interest to them.

Although the day they were asked to draw and were interviewed took place immediately after the school holidays, the students were told that they did not necessarily have to draw their last holiday, only their favorite, and what they had experienced was the best holiday. The children had access to color pencils, and effort was made to engage them as they drew by clarifying what they were drawing and why a particular object was meaningful to them. The first child turned in his drawing in six minutes and the last child, after 21 minutes. The interviews were conducted in a very informal draw-and-tell interactive session. The children's drawings were used as entry points into their thoughts and feelings about family holidays, and carefully selected prompts were used to follow up on their answers. For example, instead of using 'why' as prompts, the children were asked to 'tell me more about that' on the basis that the use of why questions on preschoolers may cause them to feel like that they have to defend their views (Gibson, 2012). The interviews were audio-recorded and then transcribed verbatim.

Data analysis

The primary data for this study are the children's drawings and these were analyzed following the guidelines provided in Kuhn's (2003) model of analysis of children's drawings. Each drawing was analyzed at first for descriptive elements of the personal (e.g. child, parents), natural (e.g. grass, sun), objective (e.g. table, football) and symbolic (e.g. arrow, musical notation). Then, each drawing was evaluated for the location (e.g. classroom), simultaneous social relations (e.g. two people doing two different activities) and together social relations (e.g. two people doing one activity together). These activities were also distinguished as either related to movement or otherwise. Finally, the drawings were analyzed using five thematic levels of (a) static display of persons or objects, (b) view of the activity, (c) presentation of a specific event, (d) display of objects admired by the child, and (e) display of abstract values which the child has a tendency for. Transcripts from the interviews were used as supplementary, supporting data, to the drawings, for the purpose of triangulation.

Findings and discussion

All of the drawings contain images typically associated with holidays – blue skies, sun, sea, trees, birds, butterflies and even rainbows. Whether or not these images were deliberately drawn because the children necessarily associated them with holidays cannot be confirmed except for butterflies, which is explained in the section on 'Holiday Activities' below. Twelve of the children drew themselves into their pictures of family holidays, and eight of the 12 drawings included other immediate family members of father, mother and brother, but also extended family members like 'Godma' (godmother), 'Ma-ma' (grandmother), 'Gong Gong' (grandfather), 'Yi-Yi' (auntie) and even 'Kakak' (the domestic maid). The inclusion of these family members does seem to reinforce the collectivistic Confucian value of family but it would be interesting to see if Western children would have drawn their favorite holidays differently. Figure 1 illustrates one of such family holidays. Anthony (pseudonym used, as will be for all other respondents reported) drew himself in black surrounded by seven of his family members, in yellow. The use of colors to single himself out of the group of family members may indicate a desire for individuality, a break away from the teaching of Confucianism but can only be verified through further conversation with Anthony on the subject matter, which was unfortunately not carried out in this study. Anthony's drawing could also be indicative of the emerging rise of intergenerational mobility in

Figure 1. Anthony's drawing.

the leisure space that scholars have recently highlighted (Kluin & Lehto, 2012; Murray, 2015; Schänzel & Yeoman, 2015).

Favorite family destinations

Twelve of the drawings, and supported by the interview transcript, provided information on where young Malaysian Chinese children's favorite holidays took place. Of the 12, only two was based on nature, one was the beach and another a park. In the other 10 drawings, it was clear that young Asian children's constructions of family holidays are very much associated with themed environments as constructed by tourism corporations. Figures 2, 3, and 4 are examples of how the children's drawings of buildings depict corporate media fantasies, such as Hello Kitty theme park, LEGOLAND, Club Med and Hard Rock Hotel. Other corporate representations, as articulated by the children themselves, include the KL Tower, Sentosa Island Singapore and York Hotel Singapore. The names of these holiday sites were easily recalled by children, confirming a recent investigation that in which, at least for the children in this study, 'brand recognition and symbolism starts in early childhood and that this brand knowledge may

Figure 2. Kaylan's drawing of the trampoline at Club Med.

Figure 3. Dana's drawing of the Hard Rock Hotel.

Figure 4. Bee Bee's drawing of LEGOLAND and Hello Kitty Theme Park.

play a role in the development of materialistic tendencies' (Watkins, Aitken, Robertson, Thyne, & Williams, 2016, p. 583).

On whether these destinations were decided by them, one of the respondents replied, 'My daddy chose it', while another proclaimed, 'My mummy and daddy decided'. Even in cases where the children had older siblings of high school years, the decision on where to holiday was made by their parents. The children did not seem to begrudge their parents this decision-making however, and one respondent was observed to good-naturedly laugh about how much she wanted to go to play with a girl in the hotel's restaurant but, 'My mummy don't [won't] let me'. Although Khoo-Lattimore et al. (2015) have confirmed the influence of Confucian values on Asian parents on family vacations, the findings in this study reinforce the values of respect for and subordination to parents in young Chinese children where holiday decisions are concerned. This is not to say, however, that the parents' own decisions on destinations are not also influenced by their children's logistical needs or known wishes.

Fun holiday activities

Given the strong presence of corporate tourism brands in the children's drawings, it is not surprising that holiday fun for these children includes the consumption of manufactured tourism experiences such as swimming in hotel pools (see Figure 5 below), playing on waterslides and jumping on Club Med's trampolines (see Figure 2). The majority of the children talked about swimming pools and, when probed on reasons why, a typical answer was one such as Qiao's, 'I like the pool because they have a slide'. In fact, one child even communicated his 'hate' for the swimming pool he was at because it 'got no waterslide' and 'it's not shallow'. For the young Chinese travelers and their families, holiday accommodations which list a swimming pool as one of their facilities may not be sufficiently attractive unless they also market a wading pool and/or a water play area.

Four of the drawings contained butterflies, which are nature-based imageries that one would expect to indicate preferences for natural tourism spaces. However, upon

Figure 5. Qiao's drawing of the hotel pool.

further enquiry, the children revealed some more commercial holiday locations and associated activities. For example, one child animatedly described going to the Butterfly Park and chasing after the butterflies while another talked about flying a butterfly kite with her sister. In addition, although both activities and especially the Butterfly Park would be deemed educational and chosen by parents for that purpose, none of the children revealed any hint of academic information about butterflies, even upon the researcher's prompts of simple questions such as, what are baby butterflies called, and what do butterflies eat. Given that Asian parents focus on educational activities despite being on holiday (Khoo-Lattimore et al., 2015), touristic sites aimed at families with children could do more to ensure that learning becomes a more tangible value for parents. For example, children could be presented with a picture booklet of educational information related to the site, with activities that children can complete at the site in exchange for a pride badge, medal, trophy or certificate. It is also interesting to note that although commercial tourism advertisements often contain images of families with young children on beaches, only one child in this study described fun as being on the beach and even in this case, the fun element was not the beach but a horse that was on the beach. Similar to the butterflies, horses on Malaysian beaches have been commodified for tourist rides and photographs.

The most surprising finding is the consensus on the children's conceptualization of 'fun' within their hotel rooms. Children in this study expressed excitement at having baths, in place of showers at home (Malaysian homes do not typically have bathtubs). They were also enthusiastic about connecting rooms and chattered about going in and out of the door between the rooms occupied by them and other family members. They engaged the researcher in animated conversations about jumping on hotel beds and eating cookies and cupcakes in the 'bedroom'. These elements of holiday fun are not currently common imageries in tourism advertisements. Even parents may not necessarily realize or understand that their children saw more fun in the everyday objects such as bathtubs, beds and doors than deliberated tourism activities centered on particular site visits. This finding has never been uncovered in previous research on family tourism. It is also noteworthy that these fun elements were not represented in the children's drawings and only elicited during the interviews.

The findings have thus far highlighted that at least for young Malaysian Chinese children, tourism spaces managed by corporate tourism brands stand out better than those managed by public entities (such as national parks) in their branding and promotion strategies, through the activities they offer that parents are willing to pay for. In destinations where their governments and destination managers are motivated to make parks and public leisure spaces more attractive to residents and tourists, this finding might provide insights into whether or not public parks and rainforests can compete for a share of the family tourism market, or how they might like to be promoted. For example, accommodation spaces within national parks could be marketed for their bathtubs and bouncy beds. Similarly, natural rainforest canopies could be sold as interconnecting 'rooms' for hiding and running in and out of.

Implications and conclusion

By exploring the articulations of the youngest possible members of Malaysian Chinese families about holidays, this study has contributed to the tourism and hospitality literature in two major ways. First, the study recognizes that the western notion and meaning of leisure and vacation may not necessarily apply for many Asian families (Khoo-Lattimore et al., 2015). This is timely given the rise of Asian travelers and the associated emerging academic attention on Asians in the tourism industry. For examples, Khoo-Lattimore and Mura (2016) recently published a book addressing the knowledge gaps on Asian genders; Yang, Khoo-Lattimore, and Arcodia (2017) revealed particular risk-taking travel behaviors of Asian women who travel solo; and Springer has announced a new series of tourism books focusing on the 'different aspects of Asian tourism and its intricate economic and social-cultural trends...[that] contributes to the idea that tourism, as both phenomenon and field of study, should be more inclusive and disentangled from dominant (mainly Western) ways of knowing' (www.Springer.com/series/15382). This study contributes to the incipient research focus and explicitly considers the expressions of young Chinese children as co-creators of their family holiday experience. By doing so, this study provides a crucial platform for the developments of Asian children in tourism research.

Second, the findings have highlighted the significance of Confucian values and how they play out in the travel behavior of young Chinese children in Malaysia. Communal characteristics of Confucianism, such as respect for and subordination to older family members, were evident in the data but, interestingly, traces of individualism also showed up in some of the young respondents. In the increasing calls for tourism scholars to decolonize research on and/or by Asians (Mura & Khoo-Lattimore, 2018; Yang, Lee, & Khoo-Lattimore, 2018), this study contributes to existing theoretical frameworks investigating Chinese travelers with specific constructs for empirical testing.

Third, the benefits and reasons for using children's drawings delineated in this study will serve as a guide for future scholars considering the method. The drawings very appropriately served as an entry point to the young children's abilities to express what they want and experience during family holidays. No tourism study thus far has administered this research tool on children as young as five years old but, more importantly, the availability of their drawings allowed Chinese children, who are typically more reserved and less expressive than Western children, to communicate 'something about their family holidays in an alternative, non-linguistic symbolic language' (Gamradt, 1995, p. 741). Although it is acknowledged that the children's experiences can never be objectively and/or completely captured by the researcher, the drawings and the room for the children to talk about their drawings, enables the children to express their family holiday experiences (Poria & Timothy, 2014; Small, 2008).

Practically, this study provides insights for hotel, resort and destination managers into what young children want from their holidays. Research has already shown that product and brand messages cultivated in childhood have long-lasting effects (Connell, Brucks, & Nielsen, 2014), and holiday destinations could do well to target young travelers today as their future and repeat guests. This study shows that a number of corporate tourism bodies have been fairly successful in capturing the share of mind of the young Chinese children's market, in that children as young as five years old are able to articulate their brands and associate them with positive experiences. The children's expressions in this study also point to their abilities to understand the meaning of fun, and to retrieve notions of fun-generating products (such as water slides, bath tubs and connecting doors) and experiences from their family holidays. The marketing assumptions that underlie this can be significant and have been alluded to the long-term share of heart for businesses. Tourism and even recreation and leisure providers may do well by listening to young children who by the age of five (at least in this study) are able to recognize brands. Holiday experiences should not only be made memorable for their parents but also to the young travelers in any family group. Particularly for Asian parents who emphasize the educational value of family holidays, tourism providers need to make the intangible tangible and recommendations for how to do this have been provided in the discussion above.

Finally, the small sample size of the children in this study is acknowledged and future scholars could extend the scale of this study. In addition, the perception and experiences of other Chinese children would make an interesting comparison to those in this study, as would whether or not young Western children perceives the same fun in the tourism products expressed by these Malaysian Chinese children. It should also be noted that, in this paper, the heterogeneity of Asian and Western travelers has not been elucidated, and future scholars should note the cautions that have been raised in the literature about the potential problems of treating all Asians as a homogeneous market segment (Khoo-Lattimore & Prayag, 2015; March, 1997). Relatedly, this paper has not considered the many debates on the division of epistemologies into 'Western' and non-Western' (for example, Chambers & Buzinde, 2015; Mura & Khoo-Lattimore, 2018; Tucker & Zhang, 2016) and positioning the paper from this theoretical viewpoint might reveal different findings.

Acknowledgements

The author acknowledges the principal role of Ms. Ben Li Cheah during the data collection process.

Disclosure statement

No potential conflict of interest was reported by the author.

Funding

This research was funded by Taylor's Research Grant Scheme Programme [Project Code TRGS/1/2012/TCHT/012];

ORCID

Catheryn Khoo-Lattimore http://orcid.org/0000-0003-2858-870X

References

Ap, J. (2004). Intercultural behaviour: Glimpses of leisure from an Asian perspective. In K. Weiermair & C. Mathies (Eds.), *The tourism and leisure industry: Shaping the future* (pp. 123–134). New York: Haworth Hospitality Press.

Barlow, C. M., Jolley, R. P., & Hallam, J. L. (2011). Drawings as memory aids: Optimising the drawing method to facilitate young children's recall. *Applied Cognitive Psychology*, *25*(3), 480–487.

Blichfeldt, B. S., Pedersen, B. M., Johansen, A., & Hansen, L. (2011). Tweens on holidays. In-situ decision-making from children's perspective. *Scandinavian Journal of Hospitality and Tourism*, *11*(2), 135–149.

Canosa, A, Wilson, E, & Graham, A. (2017). Empowering young people through participatory film: a postmethodological approach. *Current Issues in Tourism*, *20*(8), 894-907.

Carr, N. (2006). A comparison of adolescents' and parents' holiday motivations and desires. *Tourism and Hospitality Research*, *6*(2), 129–142.

Carr, N. (2011). *Children's and families' holiday experience*. Oxon, UK: Routledge.

Chambers, D., & Buzinde, C. (2015). Tourism and decolonisation: Locating research and self. *Annals of Tourism Research*, *51*, 1–16.

Chen, Y.-S., Lehto, X., Behnke, C., & Tang, C.-H. (2015). Investigating children's role in family dining-out choices: Evidence from a casual dining restaurant. *Journal of Hospitality Marketing & Management*, 1–20. doi:10.1080/19368623.2016.1077368

Connell, P. M., Brucks, M., & Nielsen, J. H. (2014). How childhood advertising exposure can create biased product evaluations that persist into adulthood. *Journal of Consumer Research*, *41*(1), 119–134.

Cullingford, C. (1995). Children's attitudes to holidays overseas. *Tourism Management*, *16*(2), 121-127.

Ekiz, E. H., & Au, N. (2011). Comparing Chinese and American attitudes towards complaining. *International Journal of Contemporary Hospitality Management*, *23*(3), 327–343.

Fivush, R., & Hamond, N. R. (1990). Autobiographical memory across the preschool years: Toward reconceptualizing childhood amnesia. In R. Fivush & J. Hudson (Eds.), *Knowing and remembering in young children* (pp. 223–248). Cambridge, UK: Cambridge University Press.

Fu, X., Cai, L., & Lehto, X. (2015). A Confucian analysis of Chinese tourists' motivations. *Journal of Travel & Tourism Marketing*, *32*(3), 180–198.

Gaines, B. L., Hubbard, S. S., Witte, J. E., & O'Neill, M. A. (2004). An analysis of children's programs in the hotel and resort industry market segment. *International Journal of Hospitality & Tourism Administration, 5*(4), 85–99.

Gamradt, J. (1995). Jamaican children's representations of tourism. *Annals of Tourism Research, 22*(4), 735–762.

Gibson, J. E. (2012). Interviews and focus groups with children: Methods that match children's developing competencies. *Journal of Family Theory & Review, 4*(2), 148–159.

Gram, M. (2007). Children as co-decision makers in the family? The case of family holidays. *Young Consumers: Insight and Ideas for Responsible Marketers, 8*(1), 19–28.

Gram, M., & Therkelsen, A. (2003). *Børnefamileferie: Family Holiday, report about family holiday content, decision-making processes and perceptions of Denmark as a holiday destination.* Aalborg: Denmark.

Gross, J., & Hayne, H. (1998). Drawing facilitates children's verbal reports of emotionally laden events. *Journal of Experimental Psychology: Applied, 4*(2), 163–179.

Hilbrecht, M., Shaw, S. M., Delamere, F. M., & Havitz, M. E. (2008). Experiences, perspectives, and meanings of family vacations for children. *Leisure/Loisir, 32*(2), 541–571.

Howard, D. R., & Madrigal, R. (1990). Who makes the decision: The parent or the child? The perceived influence of parents and children on the purchase of recreation services. *Journal of Leisure Research, 22*(3), 244–258.

Hsieh, Y. J., & Scammon, D. L. (2015). The role of Chinese culture in international marketing. *Paper presented at the Proceedings of the 1993 Academy of Marketing Science (AMS) Annual Conference.* Miami beach, FL.

Johns, N, & Gyimóthy, S. (2003). Postmodern family tourism at legoland. *Scandinavian Journal Of Hospitality and Tourism, 3*(1), 3-23.

Khoo-Lattimore, C. (2015). Kids on board: Methodological challenges, concerns and clarifications when including young children's voices in tourism research. *Current Issues in Tourism, 18*(9), 845–858.

Khoo-Lattimore, C., & Mura, P. (2016). *Asian genders in tourism.* Bristol, UK: Channel View Publications.

Khoo-Lattimore, C., & Prayag, G. (2015). The girlfriend getaway market: Segmenting accommodation and service preferences. *International Journal of Hospitality Management, 45*, 99–108.

Khoo-Lattimore, C., Prayag, G., & Cheah, B. L. (2015). Kids on board: Exploring the choice process and vacation needs of Asian parents with young children in resort hotels. *Journal of Hospitality Marketing & Management.* doi:10.1080/19368623.2014.914862

Khoo-Lattimore, C., Yang, E. C. L., & Lai, M. Y. (2016). Comparing the meanings of food in different Chinese societies: The cases of Taiwan and Malaysia. *Journal of Hospitality Marketing & Management*, 1–21. doi:10.1080/19368623.2016.1156042

Kim, S. S., Choi, S., Agrusa, J., Wang, K.-C., & Kim, Y. (2010). The role of family decision makers in festival tourism. *International Journal of Hospitality Management, 29*(2), 308–318.

Kluin, J. Y., & Lehto, X. Y. (2012). Measuring family reunion travel motivations. *Annals of Tourism Research, 39*(2), 820–841.

Kuhn, P. (2003). Thematic drawing and focused, episodic interview upon the drawing: A method in order to approach to the children's point of view on movement, play and sports at school. *Forum: Qualitative Social Research, 4*(1), 1–16.

Kwek, A., & Lee, Y. S. (2010). Chinese tourists and Confucianism. *Asia Pacific Journal of Tourism Research, 15*(2), 129–141.

Lee, C. K.C., Wong, M. Y., & Brown, R. (2015). A preliminary study on the influence of the Confucian dynamism on family decision making. *Paper presented at the Proceedings of the 1997 World Marketing Congress.*

Lehto, X. Y., Choi, S., Lin, Y.-C., & MacDermid, S. M. (2009). Vacation and family functioning. *Annals of Tourism Research, 36*(3), 459–479.

Liang, Y.-W. (2013). Children's influence on purchasing tourism products via the internet: Parental power versus children's power—The social power perspective. *Journal of Travel & Tourism Marketing, 30*(7), 639–661.

Lugosi, P., Robinson, R. N., Golubovskaya, M., & Foley, L. (2016). The hospitality consumption experiences of parents and carers with children: A qualitative study of foodservice settings. *International Journal of Hospitality Management, 54*, 84–94.

Malaysian Parents Choose International Schools Over National Schools These Days – What Are The Advantages? (2015). Retrieved from http://www.malaysiandigest.com/opinion/554189-malaysian-parents-choose-international-schools-over-national-schools-these-days-what-are-the-advantages.html

March, R. (1997). Diversity in Asian outbound travel industries: A comparison between Indonesia, Thailand, Taiwan, South Korea and Japan. *International Journal of Hospitality Management, 16*(2), 231–238.

Mitchell, L. M. (2006). Child-centered? Thinking critically about children's drawings as a visual research method. *Visual Anthropology Review, 22*(1), 60–73.

Mok, C., & DeFranco, A. L. (2000). Chinese cultural values: Their implications for travel and tourism marketing. *Journal of Travel and Tourism Marketing, 8*(2), 99–114.

Mura, P., & Khoo-Lattimore, C. (2018). *Locating Asian research and selves in qualitative tourism research Asian qualitative research in tourism*. Singapore: Springer.

Mura, P., & Pahlevan Sharif, S. (2015). The crisis of the 'crisis of representation'–Mapping qualitative tourism research in Southeast Asia. *Current Issues in Tourism*, 1–17. doi:10.1080/13683500.2015.1045459

Murray, L. (2015). Age-friendly mobilities: A transdisciplinary and intergenerational perspective. *Journal of Transport & Health, 2*(2), 302–307.

Ngai, E. W., Heung, V. C., Wong, Y., & Chan, F. K. (2007). Consumer complaint behaviour of Asians and non-Asians about hotel services: An empirical analysis. *European Journal of Marketing, 41*(11/12), 1375–1391.

Nickerson, N. P., & Jurowski, C. (2001). The influence of children on vacation travel patterns. *Journal of Vacation Marketing, 7*(1), 19–30.

Poria, Y., & Timothy, D. J. (2014). Where are the children in tourism research? *Annals of Tourism Research, 47*, 93–95.

Reisinger, Y., & Turner, L. W. (2002). Cultural differences between Asian tourist markets and Australian hosts, part 1. *Journal of Travel Research, 40*(3), 295–315.

Ruff, H. A., & Lawson, K. R. (1990). Development of sustained, focused attention in young children during free play. *Developmental Psychology, 26*(1), 85–93.

Schänzel, H. A. (2012). The inclusion of fathers, children and the whole family group in tourism research on families. In H. Schänzel, I. Yeoman, & E. Backer (Eds.), *Family tourism: Multidisciplinary perspectives* (pp. 67–80). Bristol, UK: Channel View.

Schänzel, H. A., & Smith, K. A. (2014). The socialization of families away from home: Group dynamics and family functioning on holiday. *Leisure Sciences: An Interdisciplinary Journal, 36* (2), 126–143.

Schänzel, H. A., & Yeoman, I. (2015). Trends in family tourism. *Journal of Tourism Futures, 1*(2), 141–147.

Schänzel, H. A., Yeoman, I., & Backer, E. (Eds.). (2012). *Family tourism: Multidisciplinary perspectives*. Bristol, UK: Channel View.

Segumpan, R. G., Zahari, J. S. A., & Jamaluddin, M. M. (2008). Tourism among families in Northern Peninsular Malaysia. *Asia-Pacific Social Science Review, 8*(2), 129–139.

Small, J. (2008). The absence of childhood in tourism studies. *Annals of Tourism Research, 35*(3), 772–789.

Tagg, S., & Seaton, A. V. (1995). Disaggregating friends and relatives in VFR-visiting friends and relatives-tourism research: the Northern Ireland evidence 1991/1993. Journal of Tourism Studies, 6(1), 6.

Therkelsen, A. (2010). Deciding on family holidays—Role distribution and strategies in use. *Journal of Travel & Tourism Marketing, 27*(8), 765–779.

Thornton, P. R., Shaw, G., & Williams, A. M. (1997). Tourist group holiday decision-making and behaviour: The influence of children. *Tourism Management, 18*(5), 287–297.

Todd Weaver, S., Moschis, G. P., & Davis, T. (2011). Antecedents of materialism and compulsive buying: A life course study in Australia. *Australasian Marketing Journal, 19*(4), 247–256.
Tse, T. S. M., & Hobson, J. S. P. (2008). China's outbound tourism as a way of ordering. *Journal of China Tourism Research, 7*(4), 490–505.
Tucker, H., & Zhang, J. (2016). On Western-centrism and 'Chineseness' in tourism studies. *Annals of Tourism Research, 61*, 250–252.
Turley, S. K. (2001). Children and the demand for recreational experiences: The case of zoos. *Leisure Studies, 20*(1), 1–18.
UNWTO. (2016). *UNWTO tourism highlights (2016)*. Retrieved from Madrid. Madrid, Spain: UNWTO.
Wang, K.-C., Hsieh, A.-T., Yeh, Y.-C., & Tsai, C.-W. (2004). Who is the decision-maker: The parents or the child in group package tours? *Tourism Management, 25*(2), 183–194.
Watkins, L., Aitken, R., Robertson, K., Thyne, M., & Williams, J. (2016). Advertising's impact on pre-schoolers' brand knowledge and materialism. *International Journal of Consumer Studies, 40*(5), 583–591.
Weaver, D., Becken, S., Ding, P., Mackerras, C., Perdue, R., Scott, N., & Wang, Y. (2015). Research agenda for tourism and the Chinese dream dialogues and open doors. *Journal of Travel Research, 54*(5), 578–583.
Wong, E. (1996). *A study of educational value in a theme park with reference to a case study of Ocean Park.* (Bachelors Dissertation). Hong Kong Polytechnic University, Hong Kong.
Yang, E. C. L., Khoo-Lattimore, C., & Arcodia, C. (2016). A narrative review of Asian female travellers: Looking into the future through the past. *Current Issues in Tourism*, 1–20. doi:10.1080/13683500.2016.1208741
Yang, E. C. L., Khoo-Lattimore, C., & Arcodia, C. (2017). Constructing space and self through risk taking: A case of Asian solo female travelers. *Journal of Travel Research.* doi:10.1177/0047287517692447
Yang, E. C. L., Lee, J. S. H., & Khoo-Lattimore, C. (2018). Asian cultures and contemporary tourism: Locating Asia, cultural differences and trends. In C. L. Yang & C. Khoo-Lattimore (Eds.), *Asian cultures and contemporary tourism.* Singapore: Springer.

Spatial Memory Bias in Children Tourists

Xiaoting Huang, Linlin Zhang and Lucie Ihnatoliova

ABSTRACT

What do children absorb from tourism activities? Memory can provide answers to this question – but children's memories do not necessarily tell the truth. Their recollections could be blurred by time, interference, or imagination. Taking Ocean Park Hong Kong as a case study, this research was carried out to explore children tourists' spatial memory bias by matching activity paths based on GPS recordings and recall diaries via biological sequence alignment. False memories and forgetting are the two types of memory bias of interest in this paper. Findings show that false memories, compared to forgetting, are often generated from later parts of the tourism experience. The spatial distribution of tourism memory bias is influenced by the entrance effect, aggregation effect, and competition effect. These results provide new insight into research on children education and tourist memory.

儿童旅游者的空间记忆偏差

摘要

旅游记忆作为一种相对稳定的心理状态，是旅游体验对儿童旅游者得以发挥长期影响的重要载体。然而，对于仍处于智力发育阶段的儿童旅游者而言，他们获得的旅游记忆往往存在偏差，这些记忆可能随时间衰退，可能被无关刺激扭曲，也可能因儿童旅游者自身的想象变得失真。本文尝试对儿童旅游者在旅游过程中产生的遗忘和虚假记忆两种现象进行探讨。文章以香港海洋公园为案例地，借助生物序列比对的基本操作，将由GPS设备记录的活动路径与由儿童旅游者基于记忆完成的活动日志进行匹配，从遗忘和虚假记忆两个维度探讨儿童旅游者的空间记忆偏差。研究结果表明，在时间上，儿童旅游者的遗忘现象更容易发生在第十个记忆事件左右，而虚假记忆更容易出现在第二十个记忆事件左右；在空间上，儿童游客的旅游记忆偏差受入口效应， 聚集效应和竞争效应的影响。以上结果为儿童教育和游客记忆的研究提供了新的思路。

1. Introduction

Children can experience social and cognitive development, happiness, and rehabilitation through tourism (Feng & Li, 2016). Long-term memory is the foundation of children's experience accumulation and psychological development (James, 1890). Due to living in a simpler environment than adults, children are more likely to perceive travel experiences, which encompass non-routine activities, as autobiographical memories (Lin, 2009).

Autobiographical memory is a memory system of a person's life that is clearer and more enduring than other memory types (Williams, Conway, & Cohen, 2008). Additionally, children have come to play increasingly important roles in family travel decision making; because these youngsters are future tourism consumers (McNeal, 1992), learning more about children's memories of tourism activities can facilitate tourism-related product development and marketing.

Investigations of memory can be traced back millennia. In Greek mythology, Mnemosyne, the goddess of memory, was the mother of nine Muise, highlighting the importance of memory in art and science. Academic research on memory is continually evolving; however, studies focusing on memories of tourism activities remain scarce.

The extant literature has mainly considered what tourists remember or forget; however, tourists occasionally recall false memories (i.e. remembering visiting somewhere they did not) (Loftus, 1979). False memories tend to be more common in children and adolescents than adults (Lindsay, Johnson, & Kwon, 1991; Yang, 2012), making it difficult to determine how tourism experiences affect them. The aim of this paper is to learn more about children tourists' false memories and forgetting of tourism experiences.

2. Literature review

2.1. Memory bias

2.1.1. Studies of memory bias in psychology

Memory is not as reliable as one might expect. Episodic memory is widely perceived as a constructive process rather than a literal reproduction of reality. Bartlett (1932) proposed pioneering ideas in memory distortion, positing that a natural bias exists between memory and reality. Bartlett asked local college students to retell the Indian folktale 'The War of Ghosts' and noted that participants tended to forget parts of the plot and added new material to make the story more reasonable. Loftus (1979) demonstrated that people often fill gaps in memory with imagined content according to their personal traits. One of the most notable pitfalls in memory recall is source misattribution (Manning, Loftus, & Sherman, 1996), in which people encounter challenges differentiating fantasy from objective experiences. Individual knowledge, emotional states, intelligence, and experience also affect the extent of false memories (Yang, 2012). Schwartz, Fisher, and Hebert (1988) pointed out that the ratio of false memories declined as the presentation time of stimulus increased. Underwood (1965) conducted an experiment confirming that familiarity can affect participants' rates of false memories. Forgetting is another common memory-related phenomenon and has garnered wide attention. Four main theories elucidate forgetting; of these, Waugh and Norman (1965) interference theory has gained particular support. This theory suggests that loss of short-term memory chiefly results in interference from other memory items, but not decay. Numerous types of interference have been shown to influence forgetting. Newly acquired information can interfere with and impede recall of previous memories (Barnes & Underwood, 1959). Conversely, recalled events that occurred earlier can also make it more difficult to remember current material (Rundus, 1973). These phenomena are respectively referred to as retroactive interference and proactive interference. Ratcliff, Clark, and Shiffrin (1990) raised the intensity of irrelevant stimuli by extending their presentation time or increasing their number of occurrences and found that participants' recall of items without reinforcement was significantly worse. Age, mood, health status, and repetition can also influence the extent to which a person forgets (Gerrig, Zimbardo, Campbell, Cumming, & Wilkes, 2015).

2.1.2. Studies of memory bias in children

Children's memory characteristics seem to differ from those of adults, presumably due to different memory structures (Pascual-Leone & Baillargeon, 1994), memory strategies (Flavell, 1977), and levels of knowledge and experience (Bartlett, 1932). Relevant studies have shown that children are more likely than adults to recall false information. Children tend to repress unpleasant memories (Christianson & Loftus, 1987; Davis, 2005; Freud, 2001). Furthermore, children can more easily confuse imagined events and reality (Foley & Johnson, 1985) or mix up past experiences with memories of current events (Jia, 2010). Children have also been shown to be significantly likely to recall nonexistent experiences (Ceci & Bruck, 1993, 1995). Lindsay et al. (1991) concluded that youth struggle with memory misattribution even as they approach adolescence.

Several classic experiments have been conducted, adapted, and improved to explore influences on children's memory. Schneider, Gruber, Gold, and Opwis (1993) replicated and extended Chi's (1978) famous experiment on chess expertise to explore the effects of familiarity on memory performance. They found that although novice adults performed better than children on a digit span memory task, expert children's memory of chess positions was superior, highlighting the importance of knowledge and experience on memory. Louttit's (1953) experimental results revealed that children remembered a pleasant experience more accurately than an unpleasant one.

Overall, extensive research has been conducted on memory bias, providing a basic paradigm and theoretical foundation for subsequent studies. Nevertheless, most psychological experiments have been performed in laboratories, such as memory testing of words, figures, and videos; comparatively few field investigations have considered real-life experiences. This study incorporates memory-related field work into tourism research.

2.2. Tourism experience memory

According to Kahneman (2011), tourism offers a means of helping people construct stories and collect memories. Research has shown that tourism memory may influence tourists' decision making, attitudes, and behavior. Many scholars have reported that people make travel-related decisions largely based on memories of prior tourism experiences (Martin, 2010; Wirtz, Kruger, Scollon, & Diener, 2003). Barnes, Mattsson, and Sorensen (2016) found that positive experiences remembered over a longer period exerted the strongest impacts on revisit intention. Ali, Hussain, and Ragavan (2014) and Dora (2017) confirmed that favorable long-term memories of experiences encouraged destination loyalty. Hull (1990) pointed out that pleasant memories of tourism experiences can affect consumers significantly, eliciting a positive mood and sense of happiness that frequently play important roles in one's life. In a study of wildlife tourism, Ballantyne, Packer, and Sutherland (2011) discovered that memorable experiences can lead to long-term adoption of environmentally sustainable practices such as protecting the environment and maintaining biodiversity. These results suggest that tourist memory deserves more attention in academic research.

The concept of memory originally fell under the umbrella of cognitive psychology. Tourism is an experience-intensive (Barnes et al., 2016) and multisensory activity involving long stimulation, various stimuli, and uncertain circumstances; these characteristics are outside the purview of traditional psychological experiments, especially laboratory research involving limited variables. Martin (2010) pointed out that many travel experiences are unconscious and stored as pieces of stories, which can then be emphasized by stimulation or interaction. Sang (2016) stated that tourist memory refers to storage of tourism events, with tourist memory construed as episodic memory. Episodic memory is defined as a collection of past personal experiences that occurred at a particular time and place (Schacter, Gilbert, & Wegner, 2011). In a slight departure from Sang, Kim (2010) described tourist memory as a form of autobiographic memory that people construct based on their interests and interpretations; in this case, autobiographic memory comprises episodic and semantic memory. Psychologists have stated that autobiographical memories may encourage personal well-being, as individuals seem to focus more on positive past experiences than difficult ones (Diener & Biswas-Diener, 2008; Zimbardo & Boyd, 2008). In other words, autobiographical memories may represent a kind of glorified recollection. Studies of tourist memory bring fresh insight to the aforementioned research areas.

Accuracy is the focal point within the few studies of tourist memory, whereas research on false memories exists in a vacuum. The features and classification of tourists as a research subject have also been neglected. Most studies on relevant influencing factors have pertained to on-site emotion, but the reasons behind tourism memory bias are much more expansive (e.g. stimulus characteristics and stimulus modes) (Yang, 2012). Tourists' memory bias toward different attractions has been ignored as well.

2.3. GPS equipment applications in tourism research

Global positioning system (GPS) devices enable researchers to collect continuous, high-intensity, high-resolution data. GPS technology began to be used in tourism research at the end of the 20th century. Asakura and Hato (2004) analyzed the characteristics of a sumo wrestling audience's recreational activities using GPS. Inspired by biochemical DNA sequences, Shoval and Isaacson (2007) sought to cluster tourists' behavior patterns in a scenic area based on GPS data. Taczanowska et al. (2014) evaluated hiking trail usage by combining GPS tracking data and graphical methods. In a study of the effectiveness of visitor management, Kidd et al. (2015) used GPS technology to track and analyze changes in tourists' routes and which staff facilitated these changes. Reinau, Harder, and Weber (2015) studied tourist behavior paths using a combination of GPS tracking and SMS technology. Cantis, Ferrante, Kahani, and Shoval (2016) segmented cruise passengers based on their movement patterns in a destination via GPS devices and identified associations between travelers' sociodemographic characteristics and movement. Zheng, Huang, and Li (2017) tried to predict tourists' next decisions by examining GPS information. GPS technology has therefore played an important role in tourism research thanks to its accuracy and real-time performance.

3. Method

3.1. Study design

In this study, we set out to analyze children tourists' memory bias by comparing their GPS recordings and recall diaries during a trip to Ocean Park Hong Kong. The children tourists in our sample each carried a GPS device that recorded their geographical location every few seconds. After visiting the park, each child was asked to complete a recall diary, circling the areas they had visited on a map marked with the name and image of each spot. Children were also instructed to plot their travel sequence. To determine the reliability of participants' recall diaries, the children were encouraged to recall their activity path carefully.

This research was intended to explore whether children tourists could correctly recall which areas they had visited in Hong Kong Ocean Park. We focused on two kinds of memory bias, false memories and forgetting. As indicated in Table 1, in this paper, *false memories* refer to a phenomenon where children tourists mistakenly noted having visited some spots they had not actually seen; *forgetting* describes a phenomenon where children tourists failed to recall certain spots they had visited. Children could also mistake one park area for another when recalling their tourism experience, which constitutes a type of memory bias not discussed in this paper due to difficult discrimination.

Biological sequence alignment was used to match children tourists' activity paths with their GPS recordings and recall diaries. GPS recordings were transformed into activity paths via a geo-fencing algorithm. GPS-tracked activity paths and recalled activity paths were analogized as a pair of biological sequences (the transferred GPS-tracked activity path represented the original biological sequence, and the recalled activity path represented a target sequence). Veridical memory was taken as the matching criterion, false memories as the insertion criterion, and forgetting as the deletion criterion.

3.2. Data acquisition

3.2.1. Case site

Ocean Park Hong Kong was selected as the case location for this study. Ocean Park Hong Kong is a well-known regional attraction for children and parents, featuring marine and land animal exhibitions, mobile games, and large-scale performances. The park, which is in southern Hong Kong, opened in 1977 and covers 17 acres. As one of the largest theme parks in Southeast Asia, its tourist reception exceeded that of domestic theme parks in 2010; 7.6 million guests from around the world visited in the 2013/14 fiscal year. Ocean Park Hong Kong was thus a suitable setting that provided an adequate sample of children for our research purposes.

The park is divided into two sections, the Waterfront (lower land) and the Summit (higher land), as shown in Figure 1. Tourists first arrive in the Waterfront, which is next to the entrance and exit gates; the Summit is only accessible via Ocean Express or cable

Table 1. The contrast of memory and reality.

		Memory	Reality
Data souse		Recall Diaries	GPS Recording
Memory Bias	**False Memory**	Have Been to	Not Have Been to
	Forgetting	Not Have Been to	Have Been to

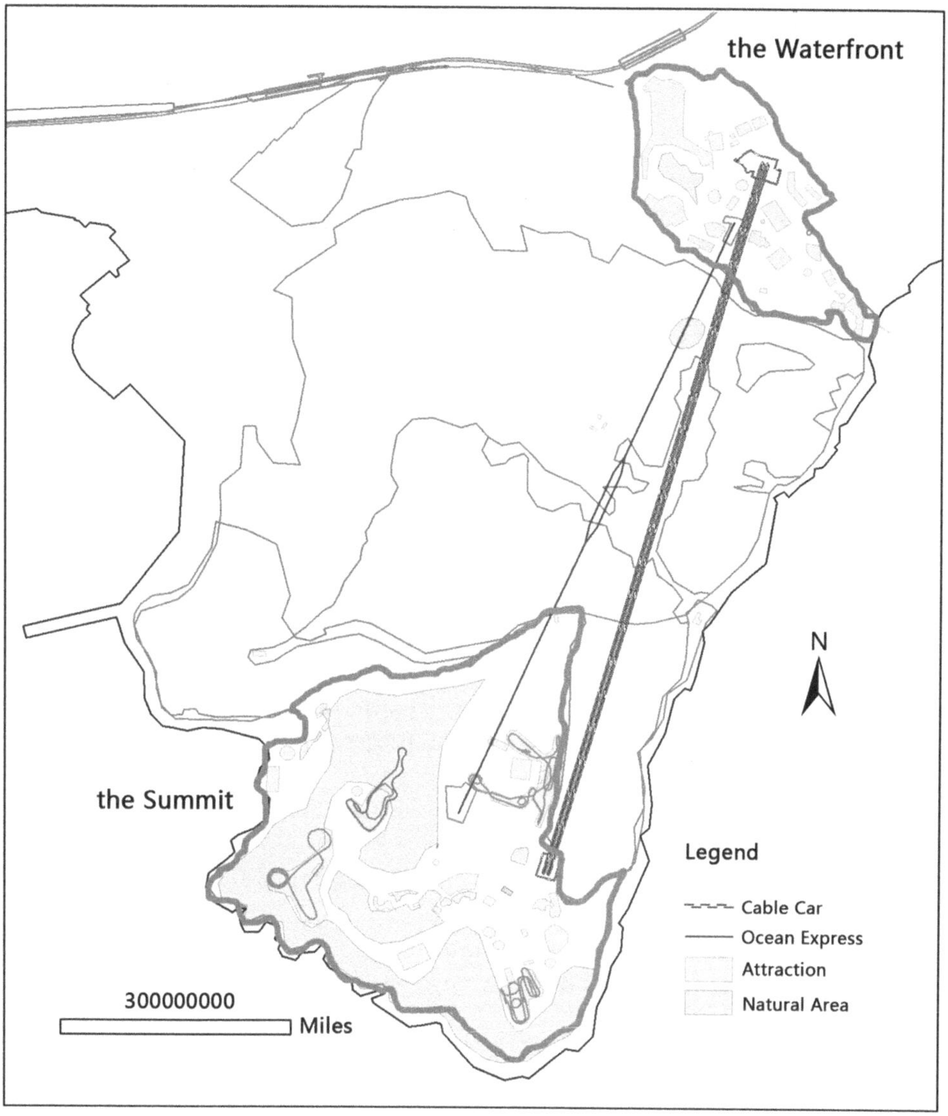

Figure 1. Distribution of facilities and attractions in Ocean Park Hong Kong.

car. When tourists finish their visit to the Summit, they return to the Waterfront and exit through the gate. We could therefore easily request the return of GPS devices and ask for children's recall dairies at the exit.

3.2.2. Sample selection and data acquisition

Data collection was carried out July 6–10, 2014. Children tourists who were younger than 18 years old and about to visit the park were randomly asked to carry a GPS device during their visit. As children were moving through the park, the GPS device recorded their geographical location every 10 seconds. When they had finished their visit, they were asked to fill out a recall diary about their activity path. Recalled activity paths were then compared to the accompanying GPS recording data from each participant.

Table 2. Sample profile.

Age			Gender		Origin		
Below 12	12–15	16–18	Female	Male	Local	China (Except HK)	International
30%	36%	34%	51%	49%	34%	57%	9%

Ultimately, 89 recall diaries matched with an uninterrupted GPS recording were collected, 62 of which passed the reliability test (i.e. the correlation between the tourist's illustrated activity path and spot list was significant). Table 2 presents the sample characteristics. Participants were roughly equally divided between boys and girls and were distributed uniformly among the following age groups: younger than 12, 12–15 years old, and 16–18 years old. Most children were from China. About one-third (34%) were local to the park, similar to the overall tourist distribution at Hong Kong Ocean Park.

3.3. Measurements

3.3.1. Geo-fencing

GPS recordings were transferred into activity paths using geo-fencing. A geo-fence is a virtual perimeter for a real-world geographic area. These fences can be dynamically generated, either as a radius around a point location or as a predefined set of boundaries (e.g. school zones or neighborhood boundaries). To determine whether each recording fell within a polygonal fence, the data transformation process included the following steps:

(1) define the geographical fence of each spot using a polygon;
(2) label the polygon using the spot code;
(3) record the time at which the tourist entered and exited this 'polygon' and then calculate the time span; and
(4) for each tourist, record the names of the polygons in chronological order if the time span exceeded 5 mins during a single visit.

3.3.2. Sequence alignment

Children tourists' activity paths based on their GPS recordings and recall diaries were matched via biological sequence alignment. Biological sequences, such as DNA, RNA, and protein sequences, are generally expressed as a string of characters from a fixed glossary. They are therefore similar to activity paths, which are composed of activities. Global alignment of activity paths was performed using the dynamic programming method and Needleman-Wunsch algorithm. The premise of this method is to insert intervals in proper locations to obtain a maximum score of the identity between two sequences. The alignment principle can be illustrated as follows:

For sequences a and b:

(1) m = length (a), n = length (b);
(2) $F(i, j)$ is the maximum score of (x_i, y_i), $0 \le i \le m$; $0 \le j \le n$;
(3) $d(x_i, y_i)$ is a kind of algorithm of (x_i, y_i);

The scoring matrix can be established as

$F(0,0) = 0$

$$F(i,j) = \max \begin{cases} 0 \\ F(i-1, j-1) + d(xi, yi) \text{ xi aligned to yi} \\ F(i-1, j) + d(xi, yi) \text{ xi aligned to a gap} \\ F(i, j-1) + d(xi, yi) \text{ yi aligned to a gap} \end{cases}$$

If a = FSHEBRDWLA, b = FSHRDCWKLA, $d(x_i = y_i) = 2$, $d(x_i \neq y_i) = -1$, then the scoring matrix will be as follows:

[2. 1. 0. 0. 0. 0. 0. 0. 0. 0]
[1. 4. 3. 2. 1. 0. 0. 0. 0. 0]
[0. 3. 6. 5. 4. 3. 2. 1. 0. 0]
[0. 2. 5. 5. 4. 6. 5. 4. 3. 2]
[0. 1. 4. 4. 4. 5. 8. 7. 6. 5]
[0. 0. 3. 3. 3. 4. 7. 7. 6. 5]
[0. 0. 2. 2. 2. 3. 6. 9. 8. 7]
[0. 0. 1. 1. 1. 2. 5. 8. 8. 7]
[0. 0. 0. 0. 0. 1. 4. 7. 10. 9]
[0. 0. 0. 0. 0. 0. 3. 6. 9. 12]

Paths based on maximum scores appear in Table 3.
The matching result is therefore
FSHEBRD-W-LA
FSH--RDCWKLA

We took biological sequence bases as activities, GPS-tracked activity paths as original sequences, and recalled activity paths as target sequences. The matching, deleting, and inserting of bases were compared to veridical memory, forgetting, and false memories, respectively (Figure 2).

To measure memory accuracy, BLAST identity was introduced to interpret the correlation between each tourists' GPS-tracked activity and recalled activity path. BLAST identity is a common measure of sequence identity. In the above alignment process, some intervals were inserted to generate more matched activities. Unmatched activities shared locations with the intervals. A tourist's activities (or activity) and interval in the same location constituted a column. We identified 12 columns containing 8 activity–activity pairs, 2 activity–interval pairs, and 2 interval–activity pairs in the above aligned paths. BLAST identity refers to the percentage of matched activities to that in the columns. For the above paths, BLAST identity was 8/12 (66.7%).

Table 3. The optimal matching path of the scoring matrix.

	F	S	H	E	B	R	D	W	L	A
F	2	1	0	0	0	0	0	0	0	0
S	1	4	3	2	1	0	0	0	0	0
H	0	3	6	5	4	3	2	1	0	0
R	0	2	5	5	4	6	5	4	3	2
D	0	1	4	4	4	5	8	7	6	5
C	0	0	3	3	3	4	7	7	6	5
W	0	0	2	2	2	3	6	9	8	7
K	0	0	1	1	1	2	5	8	8	7
L	0	0	0	0	0	1	4	7	10	9
A	0	0	0	0	0	0	3	6	9	12

```
F S H E B R D - W - L A F S H E B R D -
| | | | | | | | | | | | | | | | | | | |
F S H [-] - R D [C] W K L A F [S] H - - R D C
   forgetting  false memory   veridical memory
```

Figure 2. Veridical memory, forgetting, and false memory in activity paths.

Based on BLAST identity, we defined the false memories rate and forgetting rate as follows:

false memories rate = number of falsely recalled activities/number of recalled activities
forgetting rate = number of forgotten activities/number of visited activities
For the above paths, the false memories rate and forgetting rate were both 2/10 (20%).

4. Analysis and findings

4.1. Individual analysis

Taking Ocean Park Hong Kong as a case, internal spots were coded as shown in Table 4.

Here, the tourist ID 01 is provided as an example for illustrative purposes. We established a scoring matrix using the GPS-tracked activity path as the horizontal vector and recalled activity path as the vertical vector; the optimal matching path is depicted in Figure 3 and the matched activity path is shown in Figure 4.

Table 5 reveals the tourist's memory accuracy. This child visited 18 spots and veridically recalled 15 of them (BLAST identity = 73.50%), demonstrating strong memory accuracy overall. The forgetting rate and false memories rate were low.

Table 4. Coding table.

Spot	Code	Spot	Code
Ocean Express	TA	Hair Raiser	EA
Cable Car	TB	Bumper Blast	EB
Aqua City Lagoon	AA	Whirly Bird	EC
The Grand Aquarium	AB	The Flash	ED
Sichuan Treasures	AC	Rev Booster	EE
Sea Life carousel	AD	Bungee Trampoline	EF
Old Hong Kong	BA	Expedition Trail	FA
Gator Marsh	BB	The Rapids	FB
Panda Village	BC	Rainforest Why Zone	FC
Goldfish Treasure	BD	Sea Jelly Spectacular	GA
Amazing Bird Theater	BE	Chinese Sturgeon Aquarium	GB
Emerald Trail	BF	Flying Swing	GC
The Secret Lives of Sea horse	BG	Ferris Wheel	GD
Whisker's Theater	CA	The Dragon	GF
Merry-Go-Round	CB	Crazy Galleon	GG
Clown A Round	CC	The Abyss	GH
Balloons Up-Up-And-Away	CD	Eagle	GI
Bouncer House	CE	Veterinary Center	GJ
Frog Hopper	CF	Marine Mammal Breeding and Research Center	GK
Pinniped House	CG	Pacific Pier	GL
Toto The Loco	CH	Ocean Theater	GM
South Pole Spectacular	DA	Ocean Park Tower	GN
Arctic Fox Den	DB	Space Wheel	HA
North Pole Encounter	DC	Raging River	HB
Arctic Blast	DD	Mine Train	HC

	AB	AC	BA	BC	CH	CD	TA	DA	DB	DC	EB	EE	GC	GD	GG	GL	HC	TB
AB	1	0	0	0	0	0	0	0	0	0	0	0	0	0	0	0	0	0
AC	0	1	0	0	0	0	0	0	0	0	0	0	0	0	0	0	0	0
BA	0	0	1	0	0	0	0	0	0	0	0	0	0	0	0	0	0	0
CH	0	0	0	0	1	0	0	0	0	0	0	0	0	0	0	0	0	0
TA	0	0	0	0	0	0	1	0	0	0	0	0	0	0	0	0	0	0
DA	0	0	0	0	0	0	0	1	0	0	0	0	0	0	0	0	0	0
DB	0	0	0	0	0	0	0	0	1	0	0	0	0	0	0	0	0	0
DC	0	0	0	0	0	0	0	0	0	1	0	0	0	0	0	0	0	0
EB	0	0	0	0	0	0	0	0	0	0	1	0	0	0	0	0	0	0
EE	0	0	0	0	0	0	0	0	0	0	0	1	0	0	0	0	0	0
GD	0	0	0	0	0	0	0	0	0	0	0	0	0	1	0	0	0	0
GF	0	0	0	0	0	0	0	0	0	0	0	0	0	0	0	0	0	0
GG	0	0	0	0	0	0	0	0	0	0	0	0	0	0	1	0	0	0
GL	0	0	0	0	0	0	0	0	0	0	0	0	0	0	0	1	0	0
GH	0	0	0	0	0	0	0	0	0	0	0	0	0	0	0	0	0	0
HC	0	0	0	0	0	0	0	0	0	0	0	0	0	0	0	0	1	0
TB	0	0	0	0	0	0	0	0	0	0	0	0	0	0	0	0	0	1

Figure 3. Optimal matching path (Tourist 01).

GPS-tracked activity path	AB	AC	BA	BC	CH	CD	TA	DA	DB	DC	EB	EE	GC	GD	-	GG	GL	-	HC	TB
	I	I	I		I		I	I	I	I	I	I		I		I	I		I	I
recalled activity path	AB	AC	BA	-	CH	-	TA	DA	DB	DC	EB	EE	-	GD	GF	GG	GL	GH	HC	TB

GPS-tracked activity path: The Grand Aquarium → Sichuan Treasures → Old Hong Kong → Panda Village → Toto The Loco → Balloons Up-Up-And-Away → Ocean Express → South Pole Spectacular → Arctic Fox Den → North Pole Encounter → Bumper Blast → Rev Booster → Flying Swing → Ferris Wheel → Crazy Galleon → Pacific Pier → Mine Train → Cable Car
recalled activity path: The Grand Aquarium → Sichuan Treasures → Old Hong Kong → Toto The Loco → Ocean Express → South Pole Spectacular → Arctic Fox Den → North Pole Encounter → Bumper Blast → Rev Booster → Ferris Wheel → The Dragon → Crazy Galleon → Pacific Pier → The Abyss → Mine Train → Cable Car

Figure 4. Matched activity path (Tourist 01).

According to the temporal distribution of memory bias for Tourist 01 (Figure 5), forgetting occurred earlier and false memories emerged later. The child was less likely to mistakenly recall material from the start and end of the trip.

Table 5. The level of tourism memory bias of subject 01.

visited spots	Recalled spots	Veridically recalled spots	Blast Identity	Forgotten spots	Forgetting rate	False recalled spots	False memory rate
18	17	15	73.50%	3	16.70%	2	11.80%

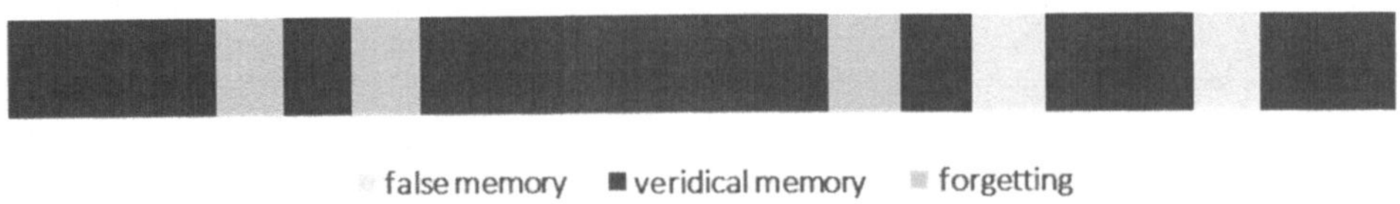

Figure 5. Temporal distribution of memory bias (Tourist 01).

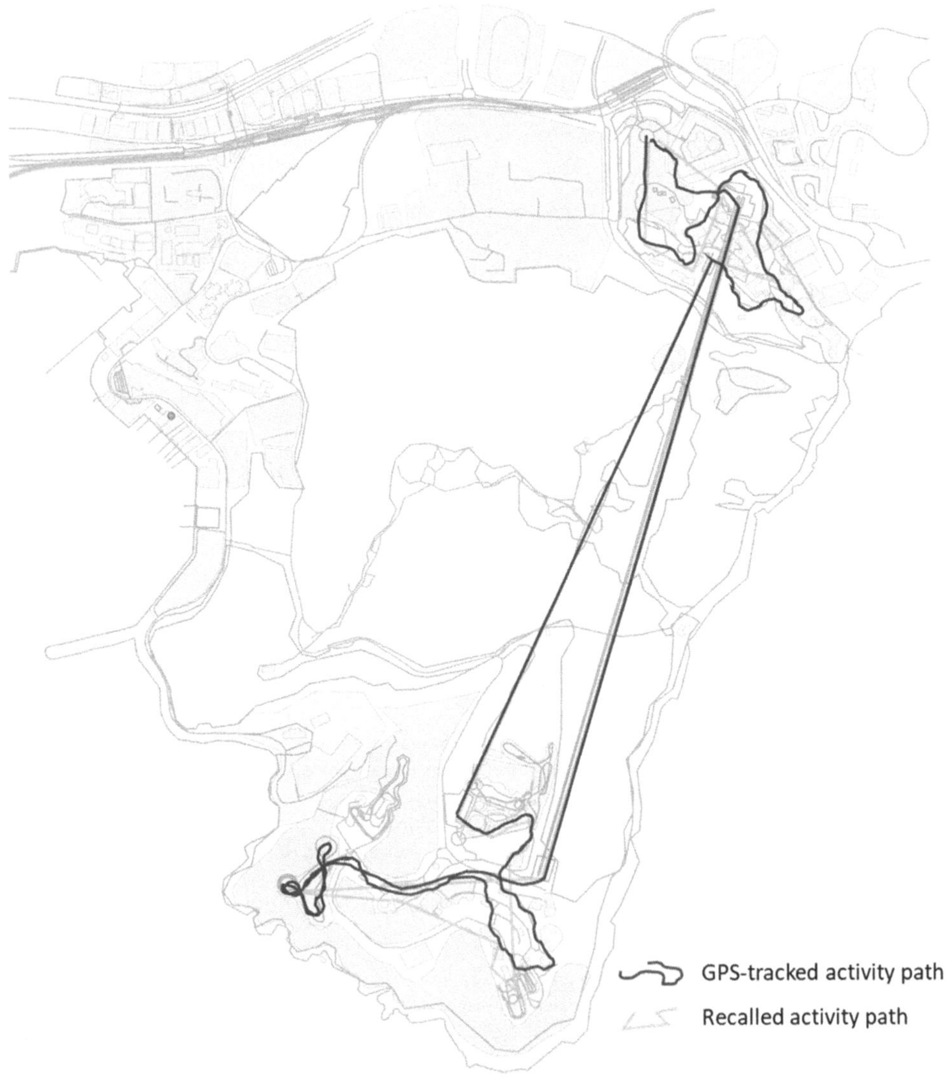

Figure 6. GPS-tracked activity path and recalled activity path (Tourist 01).

As displayed in Figure 6, Tourist 01's GPS-tracked activity path and recalled activity path overlapped heavily. Figure 7 also shows that this child mainly forgot material from the Waterfront, whereas false memories were generated exclusively in the Summit.

4.2. Integration analysis

When integrating all participants, we noticed that the level of memory bias differed greatly among the sample. As shown in Table 6, BLAST identity varied from 12.50% to 81.82%, with an average of 50.43%, which was poor. Forgetting (41.52% on average) appeared to be a more serious problem than false memories (20.75% on average).

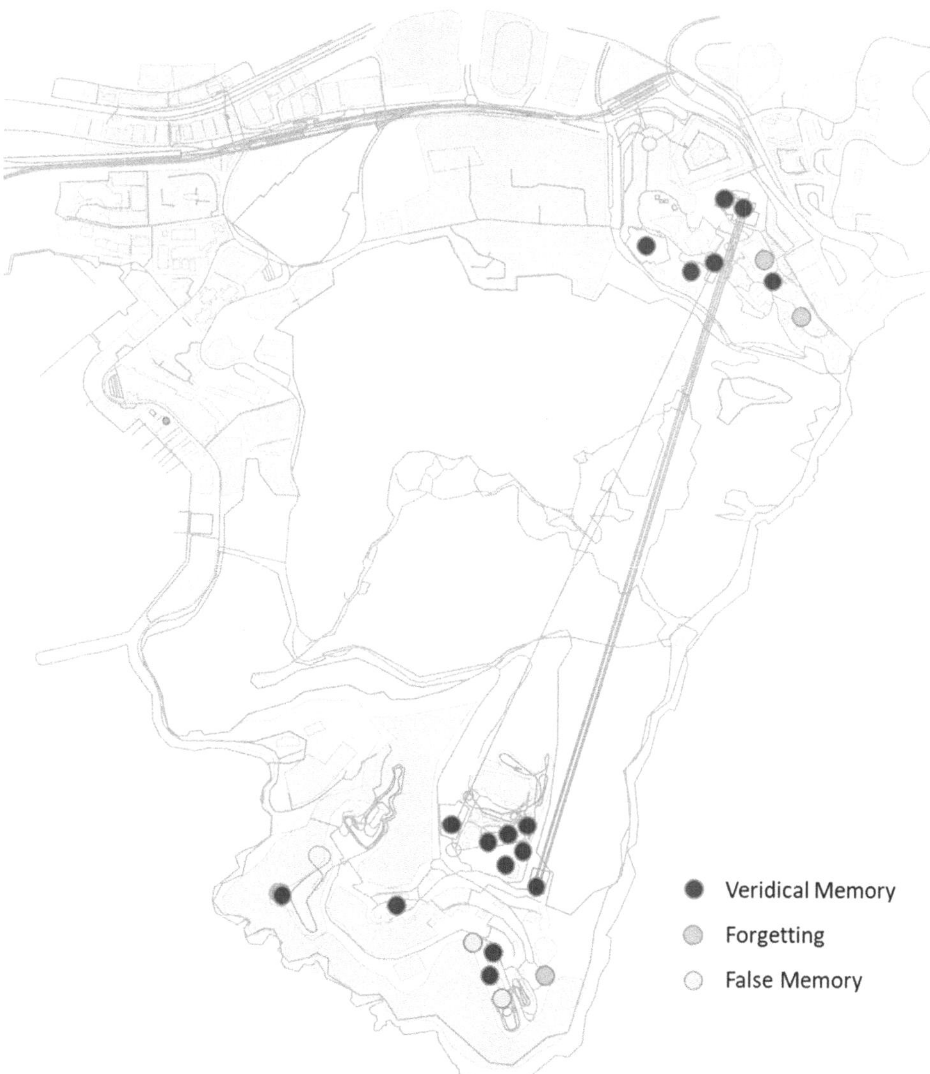

Figure 7. Spatial distribution of memory bias (Tourist 01).

Table 6. The level of tourism memory bias.

	Visited spots	Recalled spots	Veridically recalled spots	Blast identity	Forgotten spots	Forgetting rate	False recalled spots	False memory rate
Mean	17.53	12.34	9.87	50.43%	7.66	41.52%	2.47	20.75%
Std. Deviation	5.62	3.71	3.62	15.88%	4.56	18.74%	1.53	13.35%
Minimum	9	6	3	12.50%	0	0.00%	0	0.00%
Maximum	27	21	18	81.82%	19	85.71%	5	57.14%

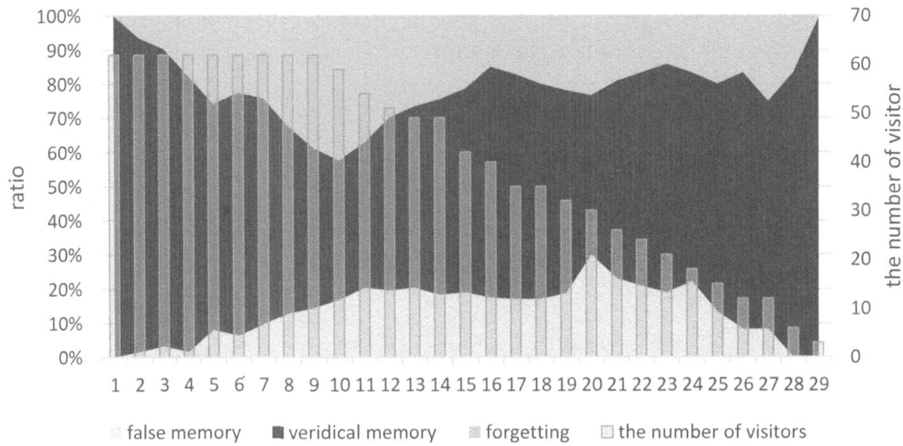

Figure 8. Temporal distribution of memory bias.

Figure 8 shows that the forgetting rate around the 10th spot visited became much higher; nearly half the children in our sample forgot the 10th place they had seen. False memories peaked around the 20th spot, although only half the children visited more than 20 places while at the park.

Figures 9 and 10 present the spatial distribution of the forgetting rate and false memories rate, respectively; a darker color denotes a higher rate. Interestingly, we noted entrance, aggregation, and competition effects in these maps. The entrance effect refers to children exhibiting a lower forgetting rate and false memories rate for spots near the entrance, consistent with the primacy effect (Barnes & Underwood, 1959) and recency effect (Rundus, 1973). The aggregation effect indicates that spots that adjoined each other and shared identical characteristics could leave an overall impression on the children. For example, the rates of forgetting and false memories for places in Polar Adventure were similar and somewhat lower than rates for other areas. The competition effect is the opposite: spots with similar characteristics could battle for a place in a child's memory. For instance, for places featuring mobile rides in Marine World, when a spot was associated with a much lower rate of forgetting or false memories, these rates were much higher for nearby places.

5. Discussion

It is not unusual for children to forget an experience or recall an experience that did not happen., In terms of time series as Figure 8 shows, children's memories of the first and

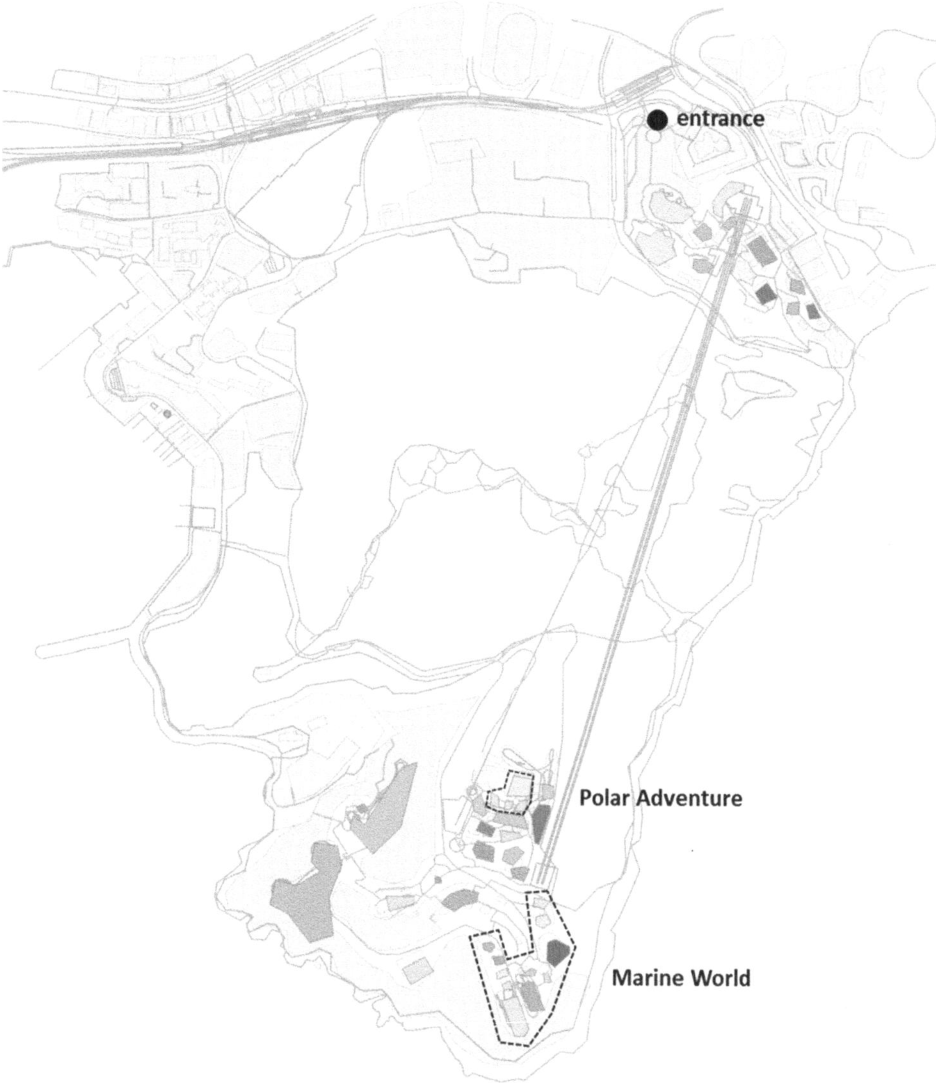

Figure 9. Spatial distribution of forgotten spots.

last few spots are obviously more accurate. children are more likely to forget the around 10th spots and generate false memories when they recall the around 20th spots. But it can't be said that forgetting occurs more often in the first half of the visit or false memory occurs more often in the second half of. As some children could merely recall about 10 spots, for whom the 10th scenic spot actually appeared in their second half.

And the meanings of space have something to do with children's forgetting and false memories. As shown in Figures 9 and 10, the aggregation effect often accompanied educational activities: South Pole Spectacular, Arctic Fox Den, and North Pole Encounter each had a similar educational theme. Yet different from the mobile rides in Marine World, forgetting and false memory rates for these three areas were all low, concurring with the levels-of-processing effect. When children expended more effort in

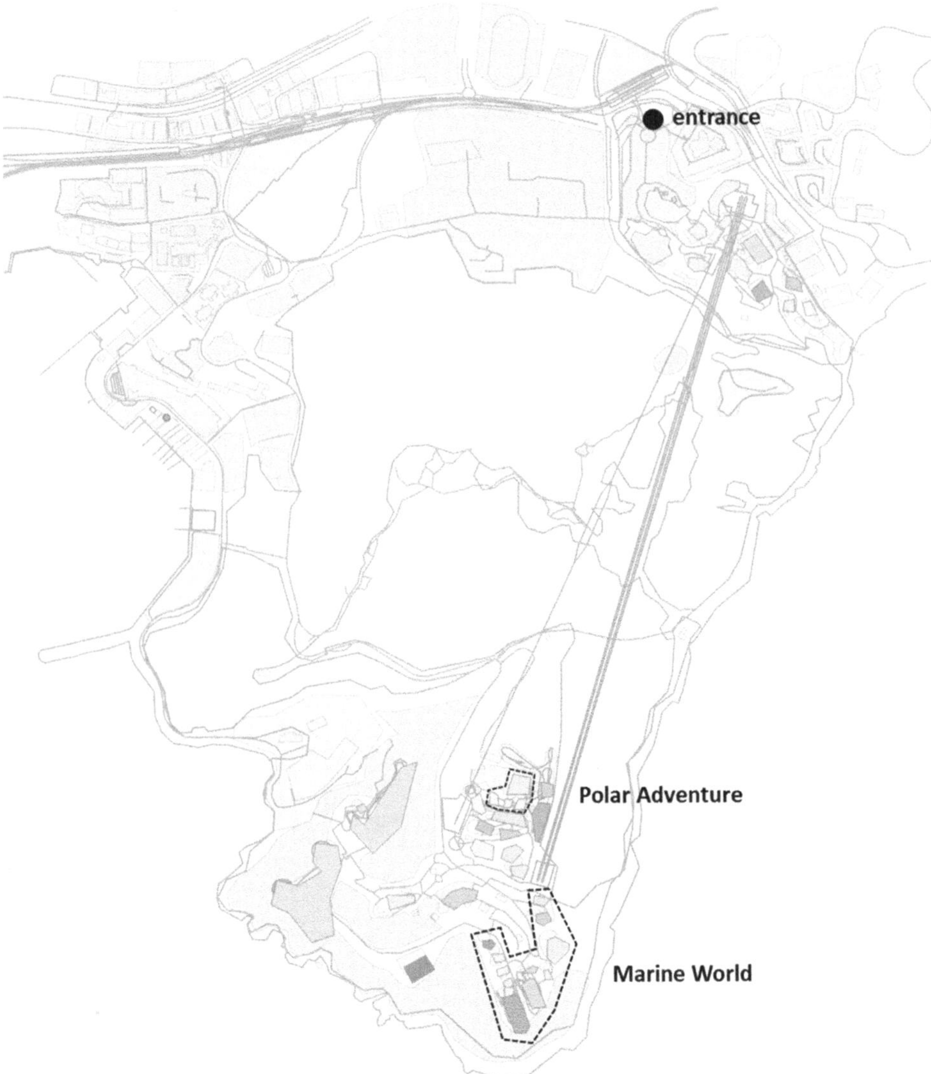

Figure 10. Spatial distribution of falsely recalled spots.

focusing on detailed information or distinctive characteristics, their visited areas could leave stronger impressions.

6. Conclusion

Children tourists' spatial memory and actual activity paths can deviate. In this study, children's GPS-tracked activity paths were compared with their recalled activity paths using biological sequence alignment. False memories and forgetting were aspects of interest; these two forms of tourism memory bias have been discussed in terms of individual, integrated, temporal, and spatial dimensions. Compared to forgetting, false memories often manifested from later tourism experiences. The entrance effect, aggregation effect, and competition effect also appeared in the spatial distribution of tourism memory bias.

For children, travel offers valuable opportunities to learn more about the world and society. Our emphasis on children's memory bias when traveling (i.e. outside of conventional educational environments) enriches theories of tourist memory and childhood education. Findings regarding children tourists' memory bias may reveal strategies for theme park developers to prevent children from forgetting a visit.

This study has several limitations. Our collected GPS data may have returned some errors due to a weak signal. Some information may therefore not have been recorded accurately or objectively. Other uncontrollable factors, such as activity path deviations during a visit, could further increase the uncertainty of our results.

Children tourists' memories have been shown to vary drastically after formation (Yang, 2012). Their travel-related recall could be consolidated, erased, or distorted in subsequent days but still become stored in long-term memory. Repeating this type of study at appropriate intervals (e.g. weeks, months, or even years after initial data collection) could help scholars trace consistencies in children tourists' memory variation. We employed the method of recognition in this investigation to help children look back on their visits within a short time; however, when children relive their memories automatically, they may not do so with support from a map or other prompt. Children's memory recall without clues appears steadier and more significant. Using an appropriate form of free recall (e.g. through interviews, written responses, or sketching) may enable researchers to collect detailed practical information from children tourists, such as familiarity with each attraction. These issues warrant further study.

Acknowledgments

I would like to extend deep gratitude to all those who have offered help and support to us. First, our thanks go to the National Natural Science Foundation for giving financial support. Second, we owe many thanks to Hong Kong Ocean Park for research sites. Third, our thanks should be sent to Zhao Ying at Sun Yat-sen University for GPS handheld devices.

Disclosure statement

No potential conflict of interest was reported by the authors.

Funding

This work was supported by the National Natural Science Foundation of China [41871138].

References

Ali, F., Hussain, K., & Ragavan, N. A. (2014). Memorable customer experience: Examining the effects of customers experience on memories and loyalty in Malaysian resort hotels. *Procedia - Social and Behavioral Sciences, 144*, 273–279.

Asakura, Y., & Hato, E. (2004). Tracking survey for individual travel behaviour using mobile communication instruments. *Transportation Research Part C: Emerging Technologies, 12*(3–4), 273–291.

Ballantyne, R., Packer, J., & Sutherland, L. A. (2011). Visitors' memories of wildlife tourism: Implications for the design of powerful interpretive experiences. *Tourism Management, 32*(4), 770–779.

Barnes, J. M., & Underwood, B. J. (1959). "Fate" of first-list associations in transfer theory. *Journal of Experimental Psychology, 58*(2), 97–105.

Barnes, S. J., Mattsson, J., & Sorensen, F. (2016). Remembered experiences and revisit intentions: A longitudinal study of safari park visitors. *Tourism Management, 57*(11), 286–294.

Bartlett, F. C. (1932). *Remembering*. Cambridge, UK: Cambridge University Press.

Cantis, S. D., Ferrante, M., Kahani, A., & Shoval, N. (2016). Cruise passengers' behavior at the destination: Investigation using GPS technology. *Tourism Management, 52*, 133–150.

Ceci, S. J., & Bruck, M. (1993). Suggestibility of the child witness: A historical review and synthesis. *Psychological Bulletin, 113*(3), 403–439.

Ceci, S. J., & Bruck, M. (1995). *Jeopardy in the courtroom: A scientific analysis of children's testimony*. Washington, DC: American Psychological Association.

Chi, M. (1978). Knowledge structure and memory development. *Children's Thinking. What Develops, 1*, 75–96.

Christianson, S., & Loftus, E. F. (1987). Memory for traumatic events. *Applied Cognitive Psychology, 1*(4), 225–239.

Davis, J. (2005). Victim narratives and victim selves: False memories syndrome and the power of accounts. *Social Problems, 52*(4), 529–548.

Diener, E., & Biswas-Diener, R. (2008). *Happiness: Unlocking the mysteries of psychological wealth*. New York: John Wiley & Sons.

Dora, A. (2017). Tourists' memories, sensory impressions and loyalty: In loco and post-visit study in Southwest Portugal. *Tourism Management, 58*, 108–118.

Feng, X., & Li, M. (2016). Children tourism: A literature review. *Tourism Tribune, 31*(9), 61–71.

Flavell, J. H. (1977). *Cognitive development*. Engiewood Cliffs: NJ: Prentice-Hall.

Foley, M., & Johnson, M. (1985). Confusions between memories for performed and imagined actions - A developmental comparison. *Child Development, 56*(5), 1145–1155.

Freud, S. (2001). *The standard edition of the complete psychological works of Sigmund Freud* (Vol. III). London: Vintage Classics.

Gerrig, R. J., Zimbardo, P. G., Campbell, A. J., Cumming, S. R., & Wilkes, F. J. (2015). *Psychology and life*. Frenchs Forest, New South Wales: Pearson Higher Education AU.

Hull, R. B. (1990). Emotion and leisure: Causes and consequences. *Journal of Leisure Research, 22*, 55–67.

James, W. (1890). *The principles of psychology*. New York: Holt.

Jia, H. (2010). A review of research on children false memories. *Studies in Early Childhood Education, 2010*(8), 47–50.

Kahneman, D. (2011). *Thinking, fast and slow*. London: Macmillan.

Kidd, A. M., Monz, C., D'Antonio, A., Manning, R. E., Reigner, N., Goonan, K. A., & Jacobi, C. (2015). The effect of minimum impact education on visitor spatial behavior in parks and protected areas: An experimental investigation using GPS-based tracking. *Journal of Environmental Management, 162*, 53–62.

Kim, J. H. (2010). Determining the factors affecting the memorable nature of travel experiences. *Journal of Travel & Tourism Marketing, 27*(8), 780–796.

Lin, C. (2009). *Developmental psychology*. Beijing: People's Education Press.

Lindsay, D., Johnson, M., & Kwon, P. (1991). Developmental changes in memory source monitoring. *Journal of Experimental Child Psychology, 52*(3), 297–318.

Loftus, E. F. (1979). The malleability of human memory: Information introduced after we view an incident can transform memory. *American Scientist, 67*(3), 312–320.

Louttit, C. M. (1953). Review of child psychology: Growth trends in psychological adjustment. *Journal of Educational Psychology, 44*(3), 187–188.

Manning, C., Loftus, E. F., & Sherman, S. J. (1996). Imagination inflation: Imagining a childhood event inflates confidence that it occurred. *Psychonomic Bulletin and Review, 3*, 208–214.

Martin, D. (2010). Uncovering unconscious memories and myths for understanding international tourism behavior. *Journal of Business Research, 63*(4), 372–383.

McNeal, J. U. (1992). *Kids as customers: Insights and implications*. Lexington, MA: Lexington Books.

Pascual-Leone, J., & Baillargeon, R. (1994). Developmental measurement of mental attention. *International Journal of Behavioral Development, 17*, 161–200.

Ratcliff, R., Clark, S. E., & Shiffrin, R. M. (1990). List-strength effect: I. Data and discussion. *Journal of Experimental Psychology: Learning, Memory, & Cognition, 16*(2), 163–178.

Reinau, K. H., Harder, H., & Weber, M. (2015). The SMS–GPS-trip method: A new method for collecting trip information in travel behavior research. *Telecommunications Policy, 39*(3–4), 363–373.

Rundus, D. (1973). Negative effects of using list items as recall cues. *Journal of Verbal Learning & Verbal Behavior, 12*(1), 43–50.

Sang, S. (2016). An exploratory study about the critical elements and memory bias in the formation process of the tourists' experience memory: Case of Chinese tourists in South Korea. *Tourism Forum, 9*(4), 33–39.

Schacter, D. L., Gilbert, D. T., & Wegner, D. M. (2011). *Semantic and episodic memory*. New York: Worth, Incorporated.

Schneider, W., Gruber, H., Gold, A., & Opwis, K. (1993). Chess expertise and memory for chess positions in children and adults. *Journal of Experimental Child Psychology, 56*(3), 328–349.

Schwartz, B. L., Fisher, R. P., & Hebert, K. S. (1988). The relation of output order and commission errors in free recall and eyewitness accounts. *Memory, 6*(3), 257–275.

Shoval, N., & Isaacson, M. (2007). Tracking tourists in the digital age. *Annals of Tourism Research, 34*, 141–159.

Taczanowska, K., González, L. M., Garciamassó, X., Muhar, A., Brandenburg, C., & Tocaherrera, J. L. (2014). Evaluating the structure and use of hiking trails in recreational areas using a mixed GPS tracking and graph theory approach. *Applied Geography, 55*, 184–192.

Underwood, B. J. (1965). False recognition produced by implicit verbal responses. *Journal of Experimental Psychology, 70*, 122–129.

Waugh, N. C., & Norman, D. A. (1965). Primary memory. *Psychological Review, 72*, 89–104.

Williams, H. L., Conway, M. A., & Cohen, G. (2008). *Autobiographical Memory*. UK: Psychology Press.

Wirtz, D., Kruger, J., Scollon, C. N., & Diener, E. (2003). What to do on spring break? The role of predicted, on-line, and remembered experience in future choice. *Psychological Science, 14*(5), 520–524.

Yang, Z. (2012). *Psychology of memory*. Shanghai: East China Normal University Press.

Zheng, W., Huang, X., & Li, Y. (2017). Understanding the tourist mobility using GPS: Where is the next place? *Tourism Management, 59*, 267–280.

Zimbardo, P., & Boyd, J. (2008). *The time paradox*. London: Rider.

What do Parents Look for in an Overseas Youth Summer Camp? Perspectives of Chinese Parents

Xinran Y. Lehto, Xiaoxiao Fu, Ksenia Kirillova and Chen Bi

ABSTRACT

Sending children to overseas summer camps has quickly become a popular practice for Chinese parents. The sizable market as a result of the potency of the Chinese economy and the current lack of understanding of parental decision-making factors are the primary motivating factors for this research. Specifically, this research attempts to shed light on the perspectives of Chinese parents concerning the specific traits and characteristics of overseas summer camps they perceive as attractive and what benefits they seek from such an experience for their children. Employing the notions of push (internal) and pull (external) factors as a theoretical backdrop, this study tested a model of perceived attractiveness of overseas youth summer camps via a sample of 234 Chinese parents. The findings suggest that both push and pull factors positively contribute to the perceived attractiveness of overseas summer camps. While sharing similar preferences with Western customers in some aspects of a summer camp, the Chinese parents were found to have unique camp attribute preferences and motives. Theoretical as well as relevant practical implications are discussed.

家长对海外青少年夏令营有何期许？基于中国家长的视角

摘要

近年来，中国家长竞相把子女送到海外参加各色夏令营。中国经济的强劲发展所带来的广袤的市场潜力，以及学术界对家长在夏令营决策上认识的不足是本研究的动机。本研究旨在探讨中国家长在为子女选择海外夏令营时偏好哪些项目特点以及追寻何种利益。运用推拉因素作为理论基础和234位中国家长作为样本，本研究调查了海外夏令营的感知吸引力。研究结果表明推与拉的因素都对海外夏令营的感知吸引力有正面影响。与传统的西方夏令营参与者不同的是，中国家长具有独特的夏令营项目偏好和参加动机。本研究为旅游学术研究和夏令营从业者提供了见解。

Introduction

Summer camps can be considered as classrooms without walls that provide expanded learning and development opportunities for children and youth. Summer camps include day camps and residential camps, with the majority of them being residential camps. Research in the domain of positive youth development has shown consistently that summer camp programs can exert positive effects on children's self-esteem, self-efficacy and other outcomes (e.g. Anshel, Muller, & Owens, 1986; Crombie, Walsh, & Trinneer, 2003; Readdick & Schaller, 2005; Yuen, Pedlar, & Mannell, 2005). As one of the most common venues for youth recreation, more than 14 million youths attend summer camps each year in the US to connect with nature, participate in various activities, and foster social development and friendships (American Camp Association, 2013). However, with increasingly global development, issues associated with diversity have started to emerge in the camp industry. These concerns include the presence of non-native language speaking campers, an influx of international staff, the difficulty of marketing camps to non-traditional camp groups, and the recognition of the diverse value systems of campers from increasingly varied backgrounds (Shelton, 2007). China is one such source market for the summer camp industry. It manifests itself as a market of tremendous growth but at the same time represents a need for research as to what this market's dispositions might be and whether existing summer camp programming and structure align with its needs.

As China's economy has soared, the Chinese middle class is rising at an astonishing rate, and is predicted to reach 76% of its population by 2022 (Iskyan, 2016). As a result, overseas camps, especially camps in the US, are becoming increasingly popular among Chinese parents aspiring to provide a better future for their children. Although different themes exist, a typical overseas summer camp experience features an organized program that offers experiential learning with both recreational and educational sources to participants (American Camp Association, 2013). Believing that education is a primary route to success, Chinese parents are willing to invest significantly in their children's education (Li, 2001). For example, more than 60,000 Chinese children crossed the Pacific for an opportunity to attend summer classes, play sports, and learn English (Cheng, 2011). Therefore, a key imperative for the camp industry is to accommodate the increased diversity of campers in order to continue to attain successful outcomes from emerging summer camp markets such as China.

The recent phenomenon of overseas camp participation from Chinese youth, however, has seen little scholarly attention. It is not known, for instance, what types of camp experiences appeal to Chinese parents and their children, and what programs and packages are most suitable for this source market. To address this knowledge gap, this research sets out to provide an understanding of camp preferences, selection criteria, and motivations from the perspective of Chinese parents. Chinese parents play a very prominent role in adolescents' consumptive choices (Carlson & Grossbart, 1988; Moschis, 1985) as China is a society where the Confucian parental authority prevails (Zhang & Carrasquillo, 1995). Owing to the single-child policy, today's Chinese parents are more preoccupied with, and tend to get more heavily involved with their children's holistic development compared with previous generations (Lehto, Fu, Li, & Zhou, 2017). Therefore, it is essential to take the parents' perspective into account when catering to China as a summer camp market. With most of the existing studies set within the Western context, research pertaining to summer camps has yet to account for the perspectives and preferences of international campers. Our research extends the existing Western perspective on summer camps and informs the summer camp industry with much needed Chinese culture-specific information. Beyond the practical objective, it was hoped that this research would contribute to the broadening of the theoretical landscape regarding summer camp.

Literature review

Push factors of overseas summer camp

Customers are thought to be motivated by both push (benefits sought) and pull (a destination's ability to deliver) factors. The push and pull framework could be used as a baseline framework to establish an understanding of the mechanism behind the perceived attractiveness of summer camps. Push factors propel a state of mind for travel (Crompton, 1979). In the context of the current research, the push factors are the benefits sought by Chinese parents for their children. Scholars have shown that parents believe that camp experiences can yield temporary pleasure and lasting memories of fun encounters, as well as significant developmental benefits (Roark & Ellis, 2009).

Previous research identified such camp experience outcomes as self-esteem and self-concept (Anshel et al., 1986; Readdick & Schaller, 2005), value and moral development (Crombie et al., 2003; Groves & Groves, 1977), family interaction (Smith, Gotlieb, Gurwitch, & Blotcky, 1987), and social capital building (Yuen et al., 2005). Exploring the components of a camp experience, Henderson, Powell, and Scanlin (2005) delineated the areas of benefits sought as positive identity, social skills as well as physical and critical thinking skills. These domains have been used in subsequent studies for the identification of summer camp benefits. The existing literature in summer camp research has also explored summer camps as avenues for youth's nutritional, physical and intellectual development (Crombie et al., 2003; Henderson et al., 2005; Ventura & Garst, 2013), sportsmanship and skill learning (Ponchillia, Armbruster, & Wiebold, 2005). Children with special needs in particular have received extensive scrutiny. For example, research has examined the outcomes of summer camps for children with severe disabilities (Rynders, Schleien, & Mustonen, 1990), children with cancer (Spirito, Forman, Ladd, & Wold, 1992), and children with chronic illness (Hunter, Rosnov, Koontz, & Roberts, 2006). Despite these efforts, there is a need to expand the scope of the existing literature and to determine the relevance of existing findings to non-Western cultural contexts, such as China.

Specifically in China, the growing economy and increasingly diversified society demand to the country's youth become well-rounded citizens. Meanwhile, employers, parents, and educators are increasingly concerned with the consequences of an exam-oriented Chinese education system (Zhao, 2014). Despite the fact that holistic student development has been the focus in China since 1999, the Chinese-style educational system still labels students by their grades. The influence of the Confucian philosophy that placed a scholar in a position of prestige coupled with the increasingly recognized need for holistic development has propelled Chinese parents to seek alternative avenues such as summer camps for their children (Chen, 2001; Lehto et al., 2017). Chinese parents believe that providing their children with overseas experiences will help their children to get ahead in the current educational system (Chen, 2001). Literature on the topic of studying abroad has also identified the significance of such benefits as cross-cultural awareness, social sensitivity, and personal development (He & Chen, 2010; Ritz, 2011). Against this background, understanding of what Chinese parents are looking for when considering overseas summer camps for their children is a timely and worthy investigation.

Pull factors of overseas summer camp

In the push–pull framework employed in tourism studies, pull factors refer to features, attractions, or attributes of a tourism destination (Crompton, 1979). These are motivational aspects that are encouraged by the attractive characteristics of a tourism destination (Yoon & Uysal, 2005). Owing to the international nature of overseas summer camps, pull factors in the context of this research include summer camp attributes pertaining to both camp destinations and the camp programs themselves.

Destination attractiveness as a topical area has received ample attention from tourism researchers. Destination attractiveness assesses the consumptive preferences and viewpoints from the demand side of tourism, and thus helps marketers understand what attributes customers are looking for in a destination (Hu & Ritchie, 1993). Consumers' preferences of destination attributes shed light on why tourists select or do not select a certain destination among many others (Vengesayi, 2003). The World Tourism Organization defined a destination as 'a physical space in which visitors spend at least one night and is made up of tourism products such as support services and attractions, and tourism resources with physical and administrative boundaries that define its management, images/perceptions of market competitiveness' (World Tourism Organization, 2003). Therefore, an overseas summer camp could be regarded as a summer destination for children.

Past literature has studied destination pull factors from the perspectives of destination resources and attractions and tended to explore the topic from the perspective of experts (Formica & Uysal, 2006; Ritchie & Zins, 1978). More recently, attention has been shifted to investigating the perspective of tourists themselves (Formica, 2002; Kim, 1998; Lee, Ou, & Huang, 2009) as they are the ultimate judges in determining the attractiveness of a region (Formica, 2002). Destination pull factors are thought to encompass a range of destination attributes. Among the most widely acknowledged aspects are tourist attractions as well as facilities and services (e.g. Hu & Ritchie, 1993; Ritchie & Zins, 1978). These two attribute dimensions could be particularly salient in the case of an international summer camp as the destination itself needs to possess certain appealing qualities in addition to providing well-developed infrastructure and hospitality services. Although not directly from a summer camp perspective, study abroad literature has provided relevant evidence in terms of what destination attributes may be appealing to overseas summer camps. Recreational opportunities, cultural and historic attractions (Eder, Smith, & Pitts, 2010; Nyaupane, Paris, & Teye, 2011; Yang, 2007), level of socio-economic development of the country (Sirakaya, Sonmez, & Choi, 2001; Um & Crompton, 1992), and the quality of the educational system (Mazzarol &

Soutar, 2002) have been identified as factors of importance when selecting a study abroad destination.

Another domain of pull factors, traditionally viewed as more important than the destination qualities, is related to summer camp programs' attributes. The features and characteristics of a product could directly affect the image of the product and thus consumers' perceptions (Chen, 2006). In research related to youth camps, Henderson et al. (2007) identified 12 camp-specific attributes that facilitated positive youth development. Among them, camp session length, staff training, staff–camper ratio, developmental activities, physical safety, staff supervision level, and cultural sensitivity and competence were found to be significantly associated with youth development outcomes. In the study abroad literature, cost, safety, reputation, quality of dining and accommodation, and staff quality are consistently found to influence the decision to go abroad to study (e.g. He & Chen, 2010; Nyaupane et al., 2011; Taylor & Rivera, 2011). Based on the insights from existing literature, it seems that summer camp attributes such as program structure (e.g. cost, duration), staff quality (e.g. staff training, staff-to-camper ratio), program image (e.g. reputation, safety), and accommodation and dining variables need to be carefully considered in assessing the preferences of the Chinese parents.

Perceived attractiveness

Perceived attractiveness is an important indicator of consumers' expectations and perceptions of a consumption experience. Tourism destination attractiveness, for example, has been defined as a combination of the relative importance of individual attributes and the perceived ability of a destination to deliver such benefits (Mayo & Jarvis, 1981). In their investigation of the antecedents of future intentions including perceived service quality, value, satisfaction, and perceived attractiveness, Um, Chon, and Ro (2006) noted that perceived attractiveness was the strongest predictor of future behavioral intentions. Perceived attractiveness is also related to perceived value, which is defined as 'the customer's overall appraisal of the net worth of the service, based on his/her assessment of what is received from the product or service and what is given (cost or sacrifice in acquiring and utilizing the service)' (Hellier, Geursen, Carr, & Rickard, 2003, p. 1765). In the context of the current study, Chinese parents' overall perceived attractiveness of overseas summer camps is important because it can lead directly to their purchase intentions of the summer camp experience for their children. It is this connection to the future intentions that presents perceived attractiveness as a concept of critical importance in targeting a market segment.

Two major approaches have been found in the literature to measure the construct of perceived attractiveness. One approach is to operationalize the construct through customer favoritism, loyalty, and future purchase intentions (Baker & Crompton, 2000; Jamal & Goode, 2001; Kim, Goh, & Yuan, 2010; Taylor, 1998). Examples of the measurements include 'intention to revisit,' 'saying positive things,' 'recommend to others,' and 'willingness to pay more.' In other instances, attitudinal measures based upon beliefs about product-specific attributes and their relative importance are used to assess the perceived attractiveness of a destination or a product (Bass & Talarzyk, 1972; Jin, 2010). Examples of the measurements include 'good,' 'favorable,' 'positive,' and 'likely'. This study combined the approaches of both behavior intentions and attitudes to evaluate the overall perceived attractiveness of overseas summer camps.

Study objectives

This study intended to explore the following research questions:

(1) What is the perceived attractiveness of overseas summer camps to Chinese parents?
(2) What are the benefits that are sought by Chinese parents who send their children to overseas summer camps?
(3) What are the summer-camp-specific attributes that Chinese parents perceive as important?
(4) Which perceived benefits and attributes contribute the most in determining Chinese parents' perceptions of attractiveness of overseas summer camps for children?
(5) Are there differences between parents who have had prior experiences of sending their children to overseas camps and those who have no prior experiences?

Based on the previous literature, the current research proposed the conceptual model in Figure 1 for assessing the Chinese youth summer camp market.

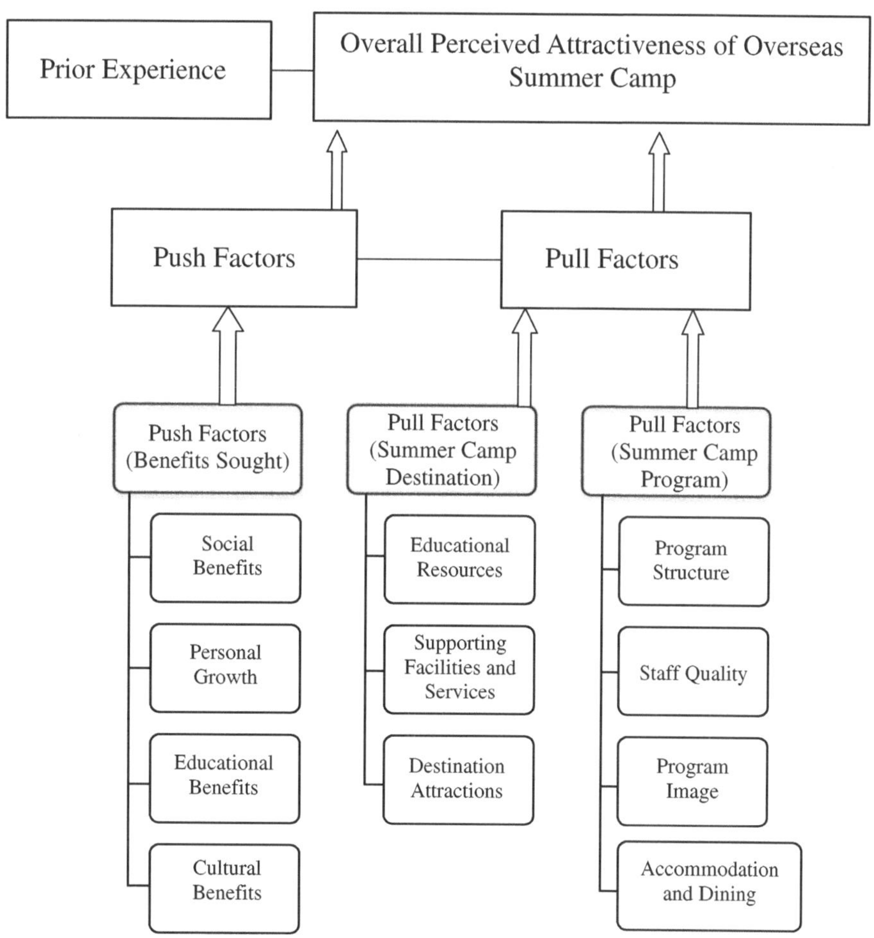

Figure 1. The conceptual model.

Method

Data collection

In this research, we aimed to investigate Chinese parents' perspectives on overseas youth summer camps, and therefore Chinese parents were the population of interest. Cross-sectional data were gathered in a two-stage process. Due to the lack of measures relevant to the potentially different non-Western market, a qualitative component was first undertaken in an attempt to identify items of interest specific to Chinese parents to supplement the existing Western-based measurement items. Sixteen in-depth semi-structured interviews were conducted with Chinese parents who sought overseas summer camps for their children. This stage of the study helped us in refining the measures utilized in the research. The quantitative component of data collection involved a survey of purposefully sampled Chinese parents. For this study, parents whose children had previously attended an overseas summer camp as well as those who expressed an interest in this summer activity for their children were included in the sample.

On-site data collection occurred in English language training centers, educational consulting agencies, and high schools in the cities of Tianjin and Beijing, China. A total of 74 questionnaires were filled out on-site. A link to an on-line questionnaire powered by Qualtrics was posted in on-line overseas camp chat groups and on the China Bulletin Board section about overseas summer camp program websites. These chat groups included parents from Beijing, Shanghai, Guangzhou, and the surrounding areas. A total of 203 responses were gathered on-line, of which 43 responses were excluded from data analysis due to substantial missing data. The final combined dataset contained 234 usable responses. T-tests did not identify significant differences between responses collected on-site and on-line.

The final sample description is presented in Table 1. The respondents were mostly college-educated parents (70%). Fifty-six percent of them were mothers with the remainder being fathers. All parents in the sample had only one child, with children's age averaging at 14 years old. Among the respondents, 60.6% had a prior experience of

Table 1. Descriptive statistics of the sample.

Demographic variable		Frequency (*N*)	Percentage (%)
Parental role	Father	102	43.6
	Mother	132	56.4
		234	
Child's gender	Male	125	53.4
	Female	109	46.6
		234	
Annual household income	Below ¥ 99,999	16	6.9
	¥ 100,000 – 199,999	52	22.3
	¥ 200,000 – 299,999	62	26.6
	¥ 300,000 – 399,999	63	27.0
	¥ 400,000 – 499,999	18	7.7
	¥ 500,000 or above	22	9.4
		233	
Education	Secondary School	6	2.6
	College/University	164	70.1
	Graduate School	64	27.4
		234	

their children attending an overseas summer camp. For parents with prior experiences, the US was the most popular choice (51.4%), followed by Australia (14.1%) and the UK (11.3%). The length of camp programs attended ranged from 6 days to 1 month, with those of three-weeks (52.1%) and two-weeks (35.2%) of length being most popular. The sampled parents obtained camp-related information from foreign language training centers (39.2%), a child's school (24.5%), and offices of educational consultants (12.6%). Similarly, camps were booked through foreign language training centers (46.2%), schools (32.2%), and educational consultants (11.2%).

Measurements

The survey instrument included socio-demographics, information about prior summer camp experiences, and information pertaining to youth summer camp push and pull factors, as well as perceived attractiveness. Fifty-six items related to the push and pull factors of summer camps as perceived by Chinese parents (see Table 2) were utilized for the purpose of this research. These items were derived based on existing literature and our qualitative interviews of Chinese parents. Following the existing research tradition, the perceived attractiveness of international summer camps was measured by six items (see Table 2). All perception items were measured on a 5-point 'Disagree–Agree' Likert scale.

Data analysis

We followed three steps for our data analysis. First, three principal component exploratory factor analyses (EFA) with varimax rotation were performed on three sets of construct measurement items respectively (push factors, camp destination attributes, and camp program attributes) in an attempt to reveal the underlying dimensions comprising Chinese parents' perceptions towards overseas youth summer camps. Second, we resorted to independent sample t-tests to explore potential differences in perception of overseas youth summer camps between parents whose children had such an experience and parents of children without prior experiences. Finally, two sets of multiple regression analyses were employed to determine the salient benefits and attributes predicting Chinese parents' perception of attractiveness of overseas youth summer camp.

Results

The means and standard deviations of measurement items are presented in Table 2. The mean scores of push attributes range from 3.42 to 4.39, with 'Allows my child to improve interpersonal communication skills' registering the highest rating. The mean values for camp destination pull attributes ranged from 3.44 to 4.19, with 'destinations with an advanced educational system' being the most highly rated. The mean values of pull attributes of summer camp programs were found to be consistently high (range 3.59–4.44), with 'safety, security, and sanitation' and 'medical and emergency system availability' rated most highly. The mean values for the items of attractiveness of overseas summer camp ranged from 3.94 to 4.28, indicating generally favorable perceptions from the respondents.

Table 2. Measurement items and their means.

Statement	Mean (s.d.)	Source
Push attributes		
Allows my child to learn a foreign language in an interactive environment	4.20 (0.62)	American Camp Association (ACA) (2005)
Allows my child to become familiar with a different educational style for studying overseas in the future	4.17 (0.74)	
Good complement to a Chinese-style education	3.80 (0.88)	
Get some experience preparing them for studying overseas in the future	4.15 (0.77)	
Allows my child to be more independent	4.07 (0.70)	
Allows my child to improve creativity and imagination	3.88 (0.77)	
Allows my child to improve judgment	3.86 (0.73)	
Allows my child to be more adventurous and try new activities	3.94 (0.68)	
Allows my child to raise his/her environmental awareness	3.52 (0.76)	Cheng (2011)
Allows my child to improve his/her decision-making skills	3.76 (0.76)	
Allows travel experiences for child's personal growth	4.33 (0.72)	
Allows my child to obtain a good visa record	3.42 (0.78)	
Allows my child to take advantage of the camp's university affiliations	3.65 (0.75)	
Allows my child to improve hands-on ability	3.68 (0.80)	
Allows my child to improve critical thinking	3.51 (0.89)	
Allows my child to build teamwork spirit	4.30 (0.55)	
Allows my child to improve interpersonal communication skills	4.39 (0.55)	
Allows my child to enjoy a different atmosphere and activities compared to those in China	4.33 (0.59)	Preliminary qualitative component
Allows my child to meet and make new friends	4.29 (0.69)	
Allows my child to increase self-esteem	3.76 (0.88)	
Allows my child to develop more leadership qualities	3.53 (0.85)	
Allows my child to be exposed to a different culture	4.36 (0.53)	
Allows my child to better accept and understand differences of nationality, race, and culture	3.80 (0.76)	
Good way of easing my child into a country and culture	4.24 (0.64)	
Pull attributes of summer camp destination		
I prefer to send my child to a country that has a high level of economic development	4.12 (0.70)	Vengesayi (2003)
I prefer to send my child to a country or region where my child may pursue a higher degree in the future	4.05 (0.71)	
I prefer to send my child to a country that has an advanced educational system that they can attend in the future	4.19 (0.63)	
I prefer to send my child to a country or region that has more natural attractions	3.81 (0.68)	
I prefer to send my child to a country or region that has more cultural attractions	3.94 (0.62)	
I prefer to send my child to a country or region that has more educational attractions	4.10 (0.63)	
I prefer to send my child to a country or region that has more historical attractions	3.85 (0.65)	
The climate of the region where the camp is located is important	3.54 (0.75)	Hu and Ritchie (1993)
Airport accessibility of the region or other regions nearby is important	3.47 (0.75)	
Entertainment and outdoor recreational opportunities are important	3.44 (0.81)	
Infrastructure of the region is important	3.83 (0.79)	
Pull attributes of summer camp programs		
Physical facilities of the camp are important	3.76 (0.73)	Henderson et al. (2005, 2007)
Accommodation quality of the camp is important	3.77 (0.60)	
Dining quality of the camp is important	3.87 (0.56)	
Staff training of the camp is important	3.89 (0.78)	
Staff-to-camper ratio of the camp is important	3.76 (0.75)	
Daily activity arrangement of the camp is important	4.14 (0.57)	
Medical and emergency system availability is important	4.38 (0.58)	
Costs of attending the camp is important	3.94 (0.69)	Vengesayi (2003)
Safety, security, and sanitation of the camp facilities are important	4.44 (0.57)	
Reputation of the organizer is important	4.27 (0.53)	
Philosophy or theme of the camp is important	4.13 (0.55)	
Group size in the camp is important	3.80 (0.63)	
Length of camp session is important	3.74 (0.63)	
Age groupings are important	3.69 (0.62)	
Diversity of campers is important	3.59 (0.69)	

(Continued)

Table 2. (Continued).

Statement	Mean (s.d.)	Source
Good personality of staff members is important	3.92 (0.78)	Preliminary qualitative component
High level of staff supervision is important	3.92 (0.78)	
Cultural competence and sensitivity of the camp are important	3.80 (0.80)	
No language barrier at the camp is important	3.86 (0.73)	
Overall camp itinerary includes complementary tour to famous universities	3.94 (0.69)	
Opportunity to experience local school life is important	4.23 (0.64)	
Overall perceived attractiveness of summer camps		
It is a good idea to send my child to overseas summer camp	4.28 (0.55)	(Baker & Crompton, 2000; Bass & Talarzyk, 1972; Kim et al., 2010)
I am in favor of sending my child to overseas summer camp	4.23 (0.57)	
I see value in sending my child to overseas summer camp	4.27 (0.55)	
It is likely that I will send my child to overseas summer camp in the near future	4.00 (0.71)	
I will recommend overseas summer camp to others	3.94 (0.72)	
I have a willingness to pay more for my child to attend overseas summer camp	3.94 (0.72)	

Scale: 1 = Strongly disagree; 2 = Disagree; 3 = Neutral; 4 = Agree; 5 = Strongly agree.

Exploratory factor analyses

The first EFA conducted on the push items originally yielded a 24-item, three-factor solution; however, 10 items were deleted due to either low loadings ($<|0.40|$) or high ($>|0.40|$) cross-loadings on two or more factors. The final three-factor solution contained 14 items, accounting for 66.67% of variance. The result of the Kaiser-Meyer-Olkin (KMO) measure of sampling adequacy indicated an acceptable level of 0.871, where a value of 0.60 or above is required for a good factor (Tabachnik & Fidell, 1996). The Bartlett's test of sphericity was also found to be significant ($p < 0.001$) (see Table 3). The resulting three factors were labeled as 'Youth development and growth', 'Social and cultural benefits', and 'Educational benefits'.

Table 3. EFA on push attributes.

Factors	F1	F2	F3
F1: Youth development and growth			
Allows my child to improve judgment	0.872		
Allows my child to improve creativity and imagination	0.853		
Allows my child to be more adventurous and try new activities	0.834		
Allows my child to be more independent	0.810		
Allows my child to improve his/her decision-making skills	0.775		
F2: Social and cultural benefits			
Allows my child to enjoy a different atmosphere and activities compared to those in China		0.904	
Allows my child to meet and make new friends		0.800	
Allows my child to improve interpersonal communication skills		0.795	
Good way of easing my child into a country and culture		0.713	
Affords cross cultural travel experiences for my child		0.663	
F3: Educational benefits			
Allows my child to become familiar with a different educational style for studying overseas in the future			0.811
Good complement to a Chinese-style education			0.781
Get some experience preparing them for studying overseas in the future			0.744
Allows my child to learn a foreign language in an interactive environment			0.677
Eigenvalue	3.793	3.463	2.745
Variance explained (%)	25.28	23.09	18.30
Cronbach's alpha	0.92	0.88	0.82

The second EFA conducted on the items of destination-related pull attributes resulted in an 11-item, three-factor solution that explained 72.27% of total variance (KMO = 0.854, Bartlett's test $p < 0.001$). The three factors are labeled as 'Supporting facilities and services', 'Educational resources', and 'Destination attractions' (Table 4). The third EFA was conducted using pull items related to camp program attributes. The initial factor solution contained three items with either low loadings (<|0.40|) or high (>|0.40|) cross-loadings on two or more factors and therefore were excluded. The subsequent EFA produced an 18-item five-factor solution, explaining 72.38% of variance (KMO = 0.867, Bartlett's test $p < 0.001$). The five resulting factors are 'Staff quality', 'Program image', 'Program structure', 'Local school visits', and 'Accommodation/dining' (Table 5).

Differences between parents with and without prior experience

A series of independent sample t-tests were conducted, and the results are presented in Table 6. It is evident that Chinese parents with prior experience consistently perceived overseas summer camps as more attractive and placed a greater importance on 'Youth development and growth', 'Social and cultural benefits', and 'Educational benefits' than parents without prior experience. In their turn, parents with no prior experience tended to view 'Staff quality' as more important. This result is understandable, given the level of anxiety and worry parents may have when sending a child overseas for the first time. Both groups viewed the availability of educational resources as very important.

Multiple regression analyses

Two multiple regression analyses were employed to identify the benefits and the attributes predicting overall perceived attractiveness of overseas summer camps by Chinese parents. The first regression model with the three benefit sought factors as predictors was significant ($F = 57.171$, $p < 0.01$), predicting 42.8% of variance in overall

Table 4. EFA on pull attributes of summer camp destinations.

Factors	F1	F2	F3
Supporting facilities and services			
Airport accessibility of the region or other regions nearby is important	0.810		
Entertainment and outdoor recreational opportunities are important	0.798		
Infrastructure of the region is important	0.740		
The climate of the region where the camp is located is important	0.737		
Educational resources			
I prefer to send my child to a country or region where my child may pursue a higher degree in the future		0.825	
I prefer to send my child to a country that has an advanced educational system that they can attend in the future		0.805	
I prefer to send my child to a country or region that has more educational attractions		0.722	
I prefer to send my child to a country that has a high level of economic development		0.631	
Destination attractions			
I prefer to send my child to a country or region that has more natural attractions			0.834
I prefer to send my child to a country or region that has more historical attractions			0.806
I prefer to send my child to a country or region that has more cultural attractions			0.782
Eigenvalue	2.71	2.66	2.58
Variance explained (%)	24.64	24.19	23.45
Cronbach's alpha	0.85	0.83	0.86

Table 5. EFA on pull items of summer camp programs.

Factors	F1	F2	F3	F4	F5
Staff quality					
High level of staff supervision is important	0.778				
Good personality of staff members is important	0.779				
Staff training of the camp is important	0.777				
Cultural competence and sensitivity of the camp are important	0.794				
Staff-to-camper ratio is important	0.753				
Daily activity arrangement of the camp is important	0.573				
Program image					
Reputation of the organizer is important		0.531			
Philosophy or theme of the camp is important		0.549			
Medical and emergency system availability is important		0.583			
Safety, security, and sanitation of the camp facilities are important		0.569			
Program structure					
Length of camp session is important			0.829		
Age groupings are important			0.760		
Group size in the camp is important			0.687		
Costs of attending the camp is important			0.636		
Local school visits					
Overall camp itinerary includes complementary tour to famous universities				0.783	
Opportunity to experience local school life is important				0.722	
Accommodation/dining					
Accommodation quality of the camp is important					0.850
Dining quality of the camp is important					0.838
Eigenvalue	4.08	2.56	2.23	1.95	1.89
Variance explained (%)	22.65	14.20	14.18	10.82	10.52
Cronbach's alpha	0.90	0.84	0.80	0.71	0.82

Table 6. Parents with prior *vs.* no prior experience of sending children to overseas summer camps.

Attribute dimensions	Prior experience (mean)	No prior experience (mean)	Mean difference
Youth development and growth	3.98	3.79	0.363**
Social and cultural benefits	4.41	4.19	0.191*
Educational benefits	4.11	4.01	0.045*
Supporting facilities and services	3.49	3.62	−0.343
Educational resources	4.13	4.06	0.081
Destination attractions	3.89	3.79	0.331*
Staff quality	3.81	4.01	−0.446**
Program image	4.32	4.26	0.208
Program structure	3.82	3.73	0.215
Local school visits	4.09	4.07	0.076
Accommodation/dining	3.82	3.78	0.119
Overall perceived attractiveness	**4.20**	**3.95**	**0.206****

A summated mean procedure was used to calculate the mean of each factor.
*$p < 0.05$; **$p < 0.01$

perceived attractiveness. 'Social and cultural benefits' ($t = 4.85$, $p < 0.001$), 'Educational benefits' ($t = 5.44$, $p < 0.001$), and 'Youth development and growth' ($t = 2.73$, $p = 0.007$) were found to be significant (see Table 7). Specifically, Educational benefits appears to be the most influential in predicting overall attractiveness of overseas camps ($\beta = 0.336$). The second regression model with the eight camp program factors as predictors was also significant ($F = 17.536$, $p < 0.01$), predicting 39.4% of variance in overall perceived attractiveness. Two significant predictors were Education resources ($t = 3.981$, $p < 0.001$) and Local school visits ($t = 3.728$, $p = 0.007$) (see Table 8).

Table 7. Benefits sought predicting overall attractiveness of overseas camps.

	Overall Attractiveness		
	Unstandardized coefficients		Standardized coefficients
Model 1	B	SE	β
(Constant)	0.999	0.243	
Youth development and growth	0.135***	0.050	0.157
Social and cultural benefits	0.318***	0.065	0.310
Educational benefits	0.296***	0.054	0.336
Adj. R^2	0.428		
F-value	57.171***		

*p < 0.1; **p < 0.05; ***p < 0.01

Table 8. Attributes predicting overall attractiveness of overseas camps.

	Overall attractiveness		
	Unstandardized coefficients		Standardized coefficients
Model 2	B	SE	β
(Constant)	0.809	0.313	
Supporting facilities and services	0.030	0.067	0.035
Educational resources	0.284***	0.073	0.293
Destination attractions	0.032	0.070	0.034
Staff quality	0.039	0.064	0.045
Program image	0.096	0.084	0.081
Program structure	0.105	0.071	0.099
Local school visits	0.181***	0.066	0.197
Accommodation/dining	0.050	0.069	0.049
Adj. R^2	0.371		
F-value	17.536***		

*p < 0.1; **p < 0.05; ***p < 0.01

Discussions and conclusion

Summer camp, as a supervised program within a controlled environment, can provide children with valued experiences and benefits. Although studies have been conducted in the US summer camp market (Anshel et al., 1986; Hassan, 2000; Readdick & Schaller, 2005), the phenomenon of Chinese children attending summer camp overseas has remained unaddressed. It was not known what types of experiences Chinese parents expect their children to obtain from an overseas summer camp and what summer camp attributes and programs were appealing to them. Given their unique cultural dispositions, it is very important to understand what Chinese parents want for their children from summer camps and what their selection criteria might be. This research is a timely effort in addressing that need by providing a detailed delineation of what their expectations and preferences are. Parents are the population of interest for this research due to the fact that it is the parents who provide funding and select camps and activity enrollment options (Barnett, 2008). This is particularly true when it comes to the Chinese summer camp market. Parents tend to be the dominant decision makers due to both the Confucian influence on consumer decisions and the still prevalent one-child family structure (Lehto et al., 2017). Chinese parents' evaluation of summer camp attractiveness is therefore particularly pertinent. The outcomes of the research provide insight for summer camp managers and marketers to improve their understanding of this emerging market, develop appropriate promotional strategies, and deliver relevant summer camp experiences for the Chinese youth.

Overall, the study results attest to Chinese parents' willingness to invest in their children's development (Chen, 2001; Lehto et al., 2017) particularly given their belief that overseas summer camps are a means to complement the Chinese test-oriented education style. In fact, in recent years, overseas summer camps have become the latest strategy for Chinese parents to envision a better future for their children. There has been growing media coverage about the phenomenon of Chinese youth flying across the Pacific for an opportunity to play sports, attend summer classes, and participate in other skill-based programs (Shu, 2014). As this research showed, US-style summer camps are becoming increasingly attractive for Chinese families.

What camp characteristics do Chinese parents value?

This research has identified eight attribute factors that can affect Chinese parents' perceived attractiveness of overseas youth summer camps. These factors can be understood at two levels – those related to the characteristics and structure of a summer camp program itself, and those pertaining to the environmental attributes related to the destination location of a camp. The fact that there is a significant destination domain is interesting as most Western participants tend to focus more on the programming specificities of a camp rather than camp location itself. This discovery is of course not surprising given that Chinese youth have to travel long haul for such camp activities.

In terms of camp-specific traits, this research noted five areas that are of importance to Chinese parents: (1) the parents emphasize the importance of provisions of quality camp staff; (2) they value camp staff that are professional and cross-culturally trained; (3) they value good program reputation and appropriate program mission; (4) they pay close attention to program structure such as having suitable camp group size, camper age composition, cost and camp length; and (5) they value good dining options and accommodation quality. These aspects of a camp, in some ways, are not dissimilar to Western participants' perspectives (e.g. He & Chen, 2010; Nyaupane et al., 2011; Taylor & Rivera, 2011), although Chinese parents clearly place a higher value on staff cultural sensitivity. Chinese parents have also shown that they seek camp attributes beyond what Western participants usually consider (Henderson et al., 2005). One such factor is that they value highly the possibility of the camp programming including such activities as touring and experiencing local schools and universities. This sentiment is again echoed when Chinese parents reflect on the benefits of such camp activities.

Does camp location play a role in influencing Chinese parents' consideration of where to send their children to overseas summer camp? This research shows that it does. Three camp location-related factors emerged in this study. It is noted that parents consider supporting facilities and services, local educational resources, and destination attractions when deciding where to enroll their children. For supporting facilities and services, parents consider airport accessibility to the camp site, camp location climate, camp community infrastructure as well as outdoor recreational opportunities. A second aspect of a destination factor is available conventional tourism resources such as cultural, historical and natural attractions. The parents value the opportunities for children to be able to visit interesting tourism sites during a camp. This is not surprising as Chinese parents in general tend to treat travel as a meaningful way for their children to learn (Lehto et al., 2017).

A third aspect, which is also the most significant aspect in influencing parents' judgment of camp attractiveness, has to do with camp destination's availability of local educational resources. Again, as reflected in their camp programming preferences, Chinese parents place a premium on a camp location's availability and accessibility of advanced educational resources. They prefer to send their children to a country or region with reputable higher educational institutions and advanced educational systems. They seem to want to use summer camp opportunities for their children to experience potential future university communities. This finding corroborates Mazzarol and Soutar's (2002) research in that educational resources are a tangible resource of study abroad destinations and thereby an important factor for decision-making. One can conclude that in their consideration of camp location and camp program attributes, Chinese parents appreciate the opportunities to use summer camp experiences as a means to help their children be acclimatized to possible future higher education options. In particular, agglomeration of well-known universities at the camp destination was highlighted as an important attraction for Chinese parents. In some sense, summer camp in a foreign country could be regarded as a form of study abroad which combines traditional summer camp programs with traveling abroad. While the value of study abroad in learning a new culture and complementary life perspective for the development of adolescents has been increasingly acknowledged (Anderson, Lawton, Rexeisen, & Hubbard, 2006), this research provided a much-needed understanding of what parents perceived as important when sending their children overseas for summer camp programs.

What benefits are Chinese parents seeking from overseas youth summer camps?

Parents generally believe that the camp experiences they purchase for their children will not only yield transient pleasure and lasting memories of fun encounters, but also significant developmental outcomes (Roark & Ellis, 2009). Therefore, the beneficial outcomes play a significant role in parents' decisions for which summer camp their children should attend and what camp attributes they consider as important. This research identified three areas of benefits that Chinese parents are seeking from an overseas summer camp experience for their children, pertaining to the areas of youth growth, social and cultural learning as well as educational benefits. These perceptions of beneficial outcomes point to the commonalities and differences between the international summer camp market segment and the domestic summer camp segment. As this research demonstrated, education-related benefits and attributes, including educational benefits, educational resources of the camp destination, and local school visits provided by a camp program, seem to be the most salient camp selection considerations for Chinese parents. This again is reflective of Chinese parents' general tendency of seeking educational components in leisure programs (Lehto et al., 2017). This finding suggests that educational benefits are considered as a primary driver for Chinese parents to enroll their children in overseas summer camps. One particular education benefit that Chinese parents seem to attribute to overseas summer camps is that the camp experience could not only complement the test-oriented, Chinese-style elementary and secondary education but also contribute to the future college education choices. The camps are regarded as helpful in preparing the Chinese youth for the challenges of the multi-cultural yet mostly English-speaking world, and are thus perceived as instrumental in developing their future careers. Existing literature analyzing motivations to study abroad supports this result. For example, He and Chen (2010) found that learning the language of a host country was perceived as the second most important factor in contemplating the decision to study abroad. Overseas summer camps seem valued as a prelude to a better educational and learning experience for the future for the Chinese youth.

Social and cultural benefits sought was another salient push factor. This indicates that meeting and making new friends, improving inter-personal communication skills, learning local culture and accumulating international travel experience are important expectations of Chinese parents concerning outcomes of the overseas summer camps experience for their children. As a commonly sought benefit in tourism in general, social and cultural benefits have been noted as a positive outcome of study abroad experiences. Through participating in school programs and interacting with local residents such as host families, children in such overseas programs make new friends with their foreign counterparts, learn to adjust to cultural differences, and enhance cross-cultural communication skills (Eder et al., 2010; He & Chen, 2010; Henderson et al., 2005). The cross-cultural exchanges can serve as a meaningful means for the participants to better understand a different social and cultural environment (Anderson et al., 2006; Fu, Huang, Li, & Kirillova, 2017). As such, our findings corroborate the existing literature with new evidences from the perspective of Chinese parents in the context of summer camp participation.

Implications

Theoretical implications

Adopting the marketing perspective, the current research explored overseas summer camps as a product with its own characteristics, capable of satisfying a unique set of needs and wants of Chinese parents. This view provides a glimpse into what Chinese parents consider essential for satisfaction with a summer camp experience. The results of the current study provide conceptual insights that will interest summer camp researchers. This study contributes to the literature by introducing a cross-cultural perspective to the summer camp conceptualization, given that most existing summer camp studies are generally concerned with domestic camp participants (e.g. Henderson et al., 2005; Readdick & Schaller, 2005).

In order for summer camps to appeal to international markets, especially to markets that are culturally distant, such as in the case of the Chinese youth market, summer camp marketing messages may be more effective when camp programs emphasize not only the outcomes associated with holistic youth development, but also specific benefits uniquely sought by an international source market. It seems that for Chinese parents, an overseas camp is akin to a study abroad experience in which educational considerations outweigh those of leisure and personal development, which are usually associated with the camp industry in the Western context (e.g. He & Chen, 2010; Nyaupane et al., 2011; Taylor & Rivera, 2011). Therefore, our research advocates the importance of a market-oriented approach in the program design and promotion of youth summer camp market. It would be of interest to researchers in understanding what can transpire when the traditional summer camp programs face a more global and diverse audience. The infusion of culturally distant participants can signify the need for new and creative programming approaches that simultaneously satisfy domestic and international campers. In fact, international campers may bring resources that are not previously available to camp programming. For instance, by harvesting different cultural backgrounds, camps can inject cross-culturally interactional elements that will benefit both the domestic and international youth participants.

By revealing that an overseas camp is an important form of educational tourism for Chinese parents, the study findings contribute to the literature on educational tourism. It seems a summer camp can take on an added value denotation in providing educational value that is synonymous with a study abroad program. The parents' future-oriented perspective in considering summer camps is noteworthy. The presence of well-known universities near the camp location was highly valued, showcasing the parents' desire to use the summer camp experience as a springboard for their children to embark on future higher education overseas. This educational orientation of the source country can be carefully programmed into summer camps.

Last but not least, the current research has demonstrated that the benefits sought and camp attributes perceived as important, possess interesting characteristics. The joint force consisting of pull and well as push factors affects the overall perceived attractiveness of overseas summer youth camps. Results from the current study not only explain Chinese parents' motivation to send their children to overseas summer camps but can also offer insights into the mechanism behind their satisfaction. Although the push and pull framework is not new to the tourism literature, applying this framework to understanding parents' psychological processes behind overseas summer camp decision making is valuable in that it yields fresh conceptual insights into the push-pull conceptual dynamic. Overseas summer camps can be positioned as a unique travel product that possesses two layers of pull factors: those associated with the destination where the camp is located and those pertaining to the camp programs themselves. Therefore, a summer camp can be viewed as a special tourist destination by itself, which, to some extent, bears similarities with transitory event attractions (Mill & Morrison, 2012). With this perspective, the current study contributes to the tourism literature at large.

Practical implications

This research, while exploratory, provides useful market-oriented practical implications for the camp industry. The view of a summer camp as a unique tourist product not only equips the camp industry with a better understanding, but also places a great responsibility on those practitioners whose job is to attract a sufficient number of students in order to stay competitive. Given the rise of the Chinese economy (Shu, 2014), camp marketers turning their gazes to the East to find unmet needs and wants could gain a competitive advantage in the camp industry. Being mindful of the identified push and pull factors will help camp managers develop effective marketing campaigns, informational brochures, and events. For example, in their marketing messages, camps should emphasize the aspects found in this study as influential, such as social, cultural, and educational benefits as well as the educational resources of a summer camp location. Given the highly valued culture-specific belief in the camp experience as a pragmatic means to enhance children's formal education, understanding the differences in educational values and systems could be crucial for camp marketers to attract Chinese participants.

A unique and significant characteristic of camp destinations is the educational resources of the camp location, which is one example of the important findings that the summer camp organizers and managers can take home from this research. The desire of Chinese parents to ready their children for future overseas studies should not be neglected in marketing efforts. One aspect of this issue is the presence of well-known universities near the camp location. From Chinese parents' perspective, overseas summer camps are perceived as a way to introduce their children to both foreign education styles and general lifestyles, in order for their children to be better integrated into the environment in the foreseeable future. Camp organizers and marketers, therefore, should be mindful of such an expectation, in order to develop a unique selling proposition, such as customized packages that encompass camp experiences sought by Chinese parents. It could be suggested that overseas camps should collaborate with nearby educational institutions, especially universities and high schools, to include excursions to campuses and interactive activities with local students.

Limitation and future research

The limitations of the current study should be acknowledged. First, since overseas summer camps in China is a recent phenomenon, data collection proved to be difficult and the sample size was limited when compared with the larger sample size available, for example, for study tour programs. Second, the sample of this study included parents who had prior experience of sending children to overseas summer camps; however, the age of their children may exceed the age restriction when data collection was executed. This may result in bias of future intention in participating in overseas summer camps. Third, the current study only investigated the parents' perspective. Given the increasing recognition that children are active social agents in their own right, this study encourages further examinations into the children's point of view, including their preferences, consumptive values, destination perceptions, and so on. Fourth, the current study achieved a conventional motivational understanding of parents' perceived attractiveness of overseas youth summer camps. Future studies with different research designs, such as qualitative or mixed method approaches as well as alternative theoretical framings can be conducted for a better contextual understanding of the phenomenon of overseas summer camps. Fifth, while all overseas destinations were included in this study, the differences of destination characteristics were not addressed. Future studies with a larger, more diverse sample, can further explore the psychological mechanisms behind the decision to camp overseas. The differences in camp destinations, camp themes, participants' socio-cultural characteristics, and between parents' and children's perspectives also represent meaningful avenues for future research. Last but not least, other related dynamics such as perceived value, personal transformation, and subjective well-being can be included in future studies for a greater theoretical contribution.

Disclosure statement

No potential conflict of interest was reported by the authors.

ORCID

Ksenia Kirillova http://orcid.org/0000-0002-7375-5071

References

American Camp Association. (2005). *The youth development outcomes of the camp experience (YDOCE) study*. Martinsville, IN: Author.

American Camp Association. (2013). *ACA facts and trends*. Retrieved January 11, 2017, from http://www.acacamps.org/press-room/aca-facts-trends

Anderson, P., Lawton, L., Rexeisen, R., & Hubbard, A. (2006). Short-term study abroad and intercultural sensitivity: A pilot study. *International Journal of Intercultural Relations*, *30*(4), 457–469. doi:10.1016/j.ijintrel.2005.10.004

Anshel, M., Muller, D., & Owens, V. (1986). Effect of a sports camp experience on the multidimensional self-concepts of boys. *Perceptual and Motor Skills*, *63*(2), 363–366. doi:10.2466/pms.1986.63.2.363

Baker, D., & Crompton, J. (2000). Quality, satisfaction and behavioral intentions. *Annals of Tourism Research*, *27*(3), 785–804. doi:10.1016/S0160-7383(99)00108-5

Barnett, L. (2008). Predicting youth participation in extracurricular recreational activities: Relationships with individual, parent, and family characteristics. *Journal of Part and Recreation Administration*, *26*(2), 28–60.

Bass, F., & Talarzyk, W. (1972). An attitude model for the study of brand preference. *Journal of Marketing Research*, *9*(1), 93–96. doi:10.2307/3149618

Carlson, L., & Grossbart, S. (1988). Parental style and consumer socialization of children. *Journal of Consumer Research*, *15*(1), 77–94. doi:10.1086/jcr.1988.15.issue-1

Chen, H. (2001). Parents' attitudes and expectations regarding science education: Comparisons among American, Chinese-American, and Chinese families. *Adolescence*, *36*(142), 305–313.

Chen, P.-J. (2006). The attributes, consequences, and values associated with event sport tourists' behavior: A means-end chain approach. *Event Management*, *10*(1), 1–22. doi:10.3727/152599506779364651

Cheng, E. (2011). Chinese parents turn to US summer camp. *China Daily*. Retrieved June 29, 2011, from http://usa.chinadaily.com.cn/china/2011-06/26/content_12778066.htm

Crombie, G., Walsh, J., & Trinneer, A. (2003). Positive effects of science and technology summer camps on confidence, values, and future intentions. *Canadian Journal of Counselling, 37*(4), 256–269.

Crompton, J. (1979). Motivations for pleasure vacation. *Annals of Tourism Research, 6*(4), 408–424. doi:10.1016/0160-7383(79)90004-5

Eder, J., Smith, W., & Pitts, R. (2010). Exploring factors influencing student study abroad destination choice. *Journal of Teaching in Travel & Tourism, 10*(3), 232–250. doi:10.1080/15313220.2010.503534

Formica, S. (2002). Measuring destination attractiveness: A proposed framework. *Journal of American Academy of Business, 1*(2), 350–355.

Formica, S., & Uysal, M. (2006). Destination attractiveness based on supply and demand evaluations: An analytical framework. *Journal of Travel Research, 44*(5), 418–430. doi:10.1177/0047287506286714

Fu, X., Huang, Z., Li, Q., & Kirillova, K. (2017). Dissecting Chinese adolescents' overseas educational travel experiences: Movements, representations and practices. *Current Issues in Tourism*, 1–22. doi:10.1080/13683500.2017.1293621

Groves, D., & Groves, S. (1977). Trends in camping and value development. *Journal of the Association for the Study of Perception, 12*(2), 22–28.

Hassan, S. (2000). Determinants of market competitiveness in an environmentally sustainable tourism industry. *Journal of Travel Research, 38*(2), 39–45. doi:10.1177/004728750003800305

He, N., & Chen, R. (2010). College students' perceptions and attitudes toward the selection of study abroad programs. *International Journal of Hospitality & Tourism Administration, 11*(4), 347–359. doi:10.1080/15256480.2010.518525

Hellier, P., Geursen, G., Carr, R., & Rickard, J. (2003). Customer repurchase intention: A general structural equation model. *European Journal of Marketing, 37*(11/12), 1762–1800. doi:10.1108/03090560310495456

Henderson, K., Bialeschki, M., Scanlin, M., Thurber, C., Whitaker, L., & Marsh, P. (2007). Components of camp experiences for positive youth development. *Journal of Youth Development, 1*(3), 15–26. doi:10.5195/JYD.2007.371

Henderson, K., Powell, G., & Scanlin, M. (2005). Observing outcomes in youth development: An analysis of mixed methods. *Journal of Park and Recreation Administration, 23*(4), 58–77.

Hu, Y., & Ritchie, B. J. R. (1993). Measuring destination attractiveness: A contextual approach. *Journal of Travel Research, 32*(2), 25–34. doi:10.1177/004728759303200204

Hunter, H., Rosnov, D., Koontz, D., & Roberts, M. (2006). Camping programs for children with chronic illness as a modality for recreation, treatment and evaluation: An example of a mission-based program evaluation of a diabetes camp. *Journal of Clinical Psychology in Medical Settings, 13*(1), 67–80. doi:10.1007/s10880-005-9006-3

Iskyan, K. (2016). China's middle class is exploding. *Business Insider*. Retrieved January 20, 2017, from http://www.businessinsider.com/chinas-middle-class-is-exploding-2016-8

Jamal, A., & Goode, M. (2001). Consumers and brands: A study of the impact of self- image congruence on brand preference and satisfaction. *Marketing Intelligence & Planning, 19*(7), 482–492. doi:10.1108/02634500110408286

Jin, X. (2010). *Exhibition brand preference in Mainland China: The role of relationship quality and destination attractiveness* (Unpublished doctoral dissertation). The Hong Kong Polytechnic University, Hong Kong SAR

Kim, H. (1998). Perceived attractiveness of Korean destinations. *Annals of Tourism Research, 25*(2), 340–361. doi:10.1016/S0160-7383(98)00007-3

Kim, Y., Goh, B., & Yuan, J. (2010). Development of a multi-dimensional scale for measuring food tourists' motivations. *Journal of Quality Assurance in Hospitality & Tourism, 11*(1), 56–71. doi:10.1080/15280080903520568

Lee, C., Ou, W., & Huang, H. (2009). A study of destination attractiveness through domestic visitors' perspectives: The case of Taiwan's hot springs tourism sector. *Asia Pacific Journal of Tourism Research, 14*(1), 17–38. doi:10.1080/10941660902727991

Lehto, X., Fu, X., Li, H., & Zhou, L. (2017). Vacation benefits and activities: Understanding Chinese family travelers. *Journal of Hospitality & Tourism Research, 41*(3), 301–328. doi:10.1177/1096348013515921

Li, J. (2001). Expectations of Chinese immigrant parents for their children's education: The interplay of Chinese tradition and the Canadian context. *Canadian Journal of Education, 26* (4), 477–494. doi:10.2307/1602178

Mayo, E., & Jarvis, L. (1981). *Psychology of leisure travel.* Boston: CBI Publishing.

Mazzarol, T., & Soutar, G. N. (2002). Push-pull factors influencing international student destination choice. *International Journal of Educational Management, 16*(2), 82–90.

Mill, R., & Morrison, A. (2012). *The tourism system.* Dubuque, IA: Kendall Hunt.

Moschis, G. (1985). The role consumer of family communication of socialization of children and adolescents. *Journal of Consumer Research, 11*(4), 898–913. doi:10.1086/209025

Nyaupane, G., Paris, C., & Teye, V. (2011). Study abroad motivations, destination selection and pre-trip attitude formation. *International Journal of Tourism Research, 13*(3), 205–217. doi:10.1002/jtr.811

Ponchillia, P., Armbruster, J., & Wiebold, J. (2005). The national sports education camps project: Introducing sports skills to students with visual impairments through short-term specialized instruction. *Journal of Visual Impairment & Blindness, 99*(11), 685–695.

Readdick, C. A., & Schaller, G. R. (2005). Summer camp and self-esteem of school-age inner-city children. *Perceptual and Motor Skills, 101*(1), 121–130. doi:10.2466/pms.101.1.121-130

Ritchie, B. J. R., & Zins, M. (1978). Culture as a determinant of the attractiveness of a tourist region. *Annals of Tourism Research, 5*(2), 252–267. doi:10.1016/0160-7383(78)90223-2

Ritz, A. (2011). The educational value of short-term study abroad programs as course components. *Journal of Teaching in Travel & Tourism, 11*(2), 164–178. doi:10.1080/15313220.2010.525968

Roark, M. F., & Ellis, G. D. (2009). Effect of self-determination theory-based strategies for staging recreation encounters on intrinsic motivation of youth residential campers. *Journal of Park and Recreation Administration, 27*(4), 1–16.

Rynders, J., Schleien, S., & Mustonen, T. (1990). Integrating children with severe disabilities for intensified outdoor education: Focus on feasibility. *Mental Retardation, 28*, 7–14.

Shelton, M. (2007). Managing diversity: Organizational change, part one. *2007 January/February Camping Magazine: Current Issues.* Retrieved June 2, 2011, from http://www.acacamps.org/campmag/0701shelton

Shu, J. (2014). *2014海外游学人数将超30万人群渐成多元化* [2014 Educational study abroad exceeds 300,000]. Retrieved June 17, 2017, from http://edu.china.com.cn/cgym/2014-02/25/content_31587965.htm

Sirakaya, E., Sonmez, S. F., & Choi, H.-S. (2001). Do destination images really matter? Predicting destination choices of student travellers. *Journal of Vacation Marketing, 7*(2), 125–142. doi:10.1177/135676670100700203

Smith, K., Gotlieb, S., Gurwitch, R., & Blotcky, A. (1987). Impact of a summer camp experience on daily activity and family interactions among children with cancer. *Journal of Pediatric Psychology, 12*(4), 533–542. doi:10.1093/jpepsy/12.4.533

Spirito, A., Forman, E., Ladd, R., & Wold, E. (1992). Remembrance programs at camps for children with cancer. *Journal of Psychosocial Oncology, 10*(3), 103–113. doi:10.1300/J077V10N03_07

Tabachnik, B., & Fidell, L. (1996). *Using multivariate statistics.* New York, NY: Harper Collins.

Taylor, A. (1998). Hostages of fortune: The abuse of children in care. In G. Hunt (Ed.), *Whistle blowing in the social services: Public accountability and professional practice* (pp. 41–64). London: Arnold.

Taylor, M., & Rivera, J. (2011). Understanding student interest and barrier to study abroad: An exploratory study. *Consortium Journal of Hospitality & Tourism, 15*(2), 56–72.

Um, S., Chon, K., & Ro, Y.-H. (2006). Antecedents of revisit intention. *Annals of Tourism Research, 33*(4), 1141–1158. doi:10.1016/j.annals.2006.06.003

Um, S., & Crompton, J. (1992). The roles of perceived inhibitors and facilitators in pleasure travel destination decisions. *Journal of Travel Research, 30*(3), 18–25. doi:10.1177/004728759203000303

Vengesayi, S. (2003). A conceptual model of tourism destination competitiveness and attractiveness. In *ANZMAC 2003 Conference Proceedings, Adelaide, 1-3 December 2003* (pp. 637–647). Adelaide, Australia: University of South Australia.

Ventura, A., & Garst, B. (2013). Residential summer camp: A new venue for nutrition education and physical activity promotion. *International Journal of Behavioral Nutrition and Physical Activity, 10*(1), 64–77. doi:10.1186/1479-5868-10-64

World Tourism Organization. (2003). WTO Think Tank enthusiastically reaches consensus on frameworks for tourism destination success. *WTO*. Retrieved June 30, 2011, from www.worldtourism.org/education/news/news_releases/newsrelease_thinktank_pressrelea se2html

Yang, M. (2007). What attracts mainland Chinese students to Australian higher education? *Studies in Learning, Evaluation, Innovation and Development, 4*(2), 1–12.

Yoon, Y., & Uysal, M. (2005). An examination of the effects of motivation and satisfaction on destination loyalty: A structural model. *Tourism Management, 26*(1), 45–56. doi:10.1016/j.tourman.2003.08.016

Yuen, F., Pedlar, A., & Mannell, R. (2005). Building community and social capital through children's leisure in the context of an international camp. *Journal of Leisure Research, 37*(4), 494–518.

Zhang, S., & Carrasquillo, A. (1995). Chinese parents' influence on academic performance. *New York State Association for Bilingual Education Journal, 10*, 46–53.

Zhao, Y. (2014). *Who's afraid of the big bad dragon? Why China has the best (and worst) education system in the world.* San Francisco, CA: Jossey-Bass.

Vacation Travel, Marital Satisfaction, and Subjective Wellbeing: A Chinese Perspective

Liping Cai, Saerom Wang and Yunzi Zhang

ABSTRACT

Marital satisfaction is important for preserving one's marriage and overall wellbeing. While the need to identify factors that enhance marital satisfaction is evident, contributions from the field of tourism remain scarce. This study examines several linkages between and among vacation travel, marital satisfaction, and subjective wellbeing through a thematic analysis of textual data from 22 semi-structured interviews. The results reveal that vacation travel enhances marital satisfaction and improves individuals' subjective wellbeing. The study has identified four drivers for enhanced marital satisfaction, including *Mutual Devotion, Reignited Passion, Strengthened Bond*, and *Open Communication*. These drivers are manifested through various benefits of vacation travel, some of which also underlie the three outcomes of individual subjective wellbeing that emerged from the data. These outcomes are *Social Support, Self-esteem*, and *Happiness*. This study contributes to the knowledge on the role of vacation travel in enhancing marital relationship for Chinese couples in a society characterized by the convergence of Chinese traditions and Western values. As the evolving patterns of global mobility continue to intensify cultural interactions, the findings also shed light on the relationships between and among vacation travel, marital satisfaction, and subjective wellbeing beyond the Chinese population.

度假旅行，婚姻满意度和主观幸福感：以中国视角为例

摘要

婚姻满意度对于保持一个人的婚姻和整体福祉很重要。虽然需要确定提高婚姻满意度的因素，但旅游领域研究的贡献仍然很少。本研究通过对22个半结构化访谈的文本数据进行主题分析，考察了度假旅行，婚姻满意度和主观幸福感之间的几种联系。结果表明，度假旅行可以提高婚姻满意度，改善个人的主观幸福感。该研究确定了提高婚姻满意度的四个驱动因素，包括相互奉献，重塑激情，强化身心粘合和开放式沟通。这些驱动因素体现在度假旅行的各种好处上，并且根据数据显示，其中一些也构成了个人主观幸福感的三个结果的基础。这些成果包括社会支持，自尊和幸福感。本研究有助于了解度假旅行在中国传统和西方价值观趋同的社会中，对于增强中国夫妻婚姻关系的重要作用。随着不断变化的全球人口流动模式继续加剧文化互动，这些发现也在一定程度上启发和揭示了度假旅行，婚姻满意度和非中国人口的主观幸福感之间的关系。

Introduction

The increasing divorce rate is a worldwide problem (Aghajanian & Thompson, 2013; Kennedy & Ruggles, 2014). One of the major contributors to the phenomenon is people's changing views on the value of marriage. In the second half of the last century, married couples became more independent of each other as women started to enter the workforce and men to share housework with their wives (Williams, 2003). In turn, love and companionship emerged as important factors that preserved the marriage bond (Proulx, Helms, & Buehler, 2007). Given the enhanced independence, the lack of affection is more likely to lead to divorce (Amato, 2000). Research on marital satisfaction is imperative as it plays an important role in maintaining marriage and in preserving individuals' wellbeing (Stutzer & Frey, 2006). For instance, as suggested by the marital discord model of depression (Beach, Sandeen, & O'Leary, 1990), marital dissatisfaction is a significant predictor of depressive symptoms. While studies connecting marriage and wellbeing have mostly focused on the impact of marital distress on depression, the linkage between marriage and the positive aspects of personal wellbeing is yet an evolving area of inquiry that needs further investigation (Helms, 2013).

One possible vehicle for building strong ties between married couples is vacation travel. As witnessed by a government program in Malaysia, which helped married couples rebuild their relationships, vacation travel is influential in forming a greater spousal bond (Durko & Petrick, 2013). The proposition that family vacation strengthens relationships among its members has been well embraced in extant literature (Lehto, Choi, Lin, & MacDermid, 2009). However, relatively less attention has been paid to the linkage between vacation travel and the spousal relationship, albeit the understanding in leisure literature that shared leisure activities are an effective medium to enhance marital satisfaction (Voorpostel, van der Lippe, & Gershuny, 2010). Previous studies have shown that leisure and vacation travel are similar in their psychological effect on individuals, such as enjoyment (Mannell & Iso-Ahola, 1987). Extending from this idea, it can be inferred that vacation travel could have a positive effect on spousal relationship. Yet, such an inference needs to be validated as individuals are likely to have divergent experiences under different settings due to contextual variations (Ryan, 1994). Another known benefit of vacation travel for married couples is the enhancement of their individual subjective wellbeing (Aref, 2011). However, such connection has not been sufficiently established in conjunction with their marital satisfaction.

As expressed in the Chinese belief that 'if the family lives in harmony, all affairs will prosper' (Leung, Koch, & Lu, 2002, p. 202), Chinese society in general values a harmonious relationship within a family. In the past, the conjugal happiness of a couple was socially disregarded under the influence of Confucian family ethics, according to which arranged marries were the norm (Xu & Whyte, 1990). During the twentieth century, spousal relationship became the new bedrock of Chinese families. This was in part due to the shift from an arranged marriage system to an era that started to favor freedom of choice for marriage (Chen & Li, 2007). However, the increasing divorce rate today is threatening China's dominant culture of collectivism that emphasizes group cohesiveness (Zhou, 2017). As the basic yet critical unit of solidarity and harmony for Chinese society, marriage as an institution depends on the quality of the spousal relationship in the population. Despite its importance, vacation travel as a means to strengthen Chinese married couples' relationships has not been explored to the extent that the society deserves.

With this backdrop, China is selected as a case study towards three research objectives: (a) to understand the linkages among vacation travel, marital satisfaction, and subjective wellbeing for Chinese married couples; (b) to understand the factors that explain the impact of vacation travel on their marital satisfaction; and (c) to explore the factors that influence their subjective wellbeing from vacation travel. The study's sample includes Chinese nationals who currently live in the United States as visiting scholars, students, and their spouses. This sample represents a unique demographic profile holding cultural values of both the West and the East. Born and raised in China, the Chinese values are ingrained in their cultural identity. Yet, their choice of visiting and studying in the United States indicates their tendency to accept and be influenced by Western cultural values represented by the United States. As the Chinese population at home is gradually being exposed to the Western culture (Abelmann & Kang, 2014), findings from this sample are informative about the population in general and should also bear significant implications for other countries facing a similar situation, where their own cultures are increasingly blended with those of others.

Literature review

Marital satisfaction is defined as the subjective and global assessment of a couple's marital relationship (Durodoye, 1997). While some studies have used marital quality and marital satisfaction interchangeably, the consensus is that the global evaluation of marital quality is referred to as marital satisfaction (Li & Fung, 2011). For married couples, marital satisfaction serves as the baseline requirement in maintaining individual and family wellbeing (Stack & Eshleman, 1998). Keeping marriage strong is also an important social topic, as it is one of the fundamental units of a society (Cholewinski, 2002). Therefore, it is consequential to understand the factors that enhance marital satisfaction. It should be noted that drivers for marital satisfaction are not the opposite of marital distress. There are unique factors that precede marital satisfaction (Bradbury, Fincham, & Beach, 2000), which are examined as follows.

Marital satisfaction and its drivers

Marital satisfaction has been found to be closely related to the economic benefits of marriage, one of which is joint income, which facilitates a family's economic activities (Stimpson, Wilson, & Peek, 2012). Schwartz (2005) also suggests that premium wages for married men in the mid-twentieth century in the United States served utilitarian purposes for marriage. Recognizing the role of marital satisfaction in shaping individuals' emotional states (Fincham, Beach, Harold, & Osborne, 1997), previous studies have explored various contributing factors (Stutzer & Frey, 2006). First, couples' communication characteristics are an important predictor for marital satisfaction (Caughlin, 2002). While lack of communication in marriage reduces marital satisfaction, active communication, regardless of whether a topic is significant or trivial, could enhance marital satisfaction. For instance, even when married couples talk about their dissatisfaction with and irritation about the marriage in a way that could result in marital distress, their marital satisfaction increases (Addis & Bernard, 2002).

Second, defined as 'being concerned with effectively understanding oneself and others and relating well to people' (Bar-On, Handley, & Fund, 2006, p. 18), emotional intelligence also influences marital satisfaction (Dehkordi, 2012). Specifically, the extent to which couples emotionally understand each other contributes greatly to their marital happiness (Batool & Khalid, 2012). Third, equality is another driver for marital satisfaction (Beach et al., 1990). According to equity theory, inequity is correlated with low satisfaction in a close relationship (Oliver & Swan, 1989). Such a notion has been argued to be applicable to marriage. When one spouse perceives that he or she receives less from the marriage than his or her own input, individuals exhibit low marital satisfaction (Walster, Traupmann, & Walster, 1978). Fourth, similarities between partners are also a significant predictor for high marital satisfaction (Gaunt, 2006). While some have argued that the overall similarity between partners is important, Klohnen and Mendelsohn (1998) concluded that similarities in specific aspects were more influential than others. For instance, shared values and principles about romantic relationships have been found to significantly influence marital satisfaction (Chi, Epstein, Fang, Lam, & Li, 2013).

Fifth, marital satisfaction is dependent on couple's expectations (Fletcher, Simpson, & Thomas, 2000). In supporting this argument, Campbell, Simpson, Kashy, and Rholes (2001) found that fulfilling one's expectation for the partner impacted marital satisfaction. Lastly, common goals shared by a couple to sustain a healthy marriage have been identified to influence marital satisfaction (Li & Fung, 2011). It is necessary to note that each marital goal influences marital satisfaction to a different magnitude, depending on its importance to the individual. Although the studies mentioned above are informative about specific drivers for marital satisfaction, they do not offer comprehensive insights into the relationship between these drivers and marital satisfaction in the context of vacation travel.

Leisure and marital satisfaction

While existing studies are limited in understanding the association between vacation travel and marital satisfaction, their relationship may be inferred from the influence of leisure on marital satisfaction. The proposition that a couple's leisure activities enhance their marital satisfaction has been widely accepted for some time (Sharaievska, Kim, & Stodolska, 2013). A couple's leisure time has been found to increase marital satisfaction as time spent together improves communication between the two (Johnson, Zabriskie, & Hill, 2006). Further, shared leisure time allows for increased physical intimacy through more touching and hugging (Finucane & Horvath, 2000). Another assumption behind the positive link between leisure and marital satisfaction is that leisure companionship is inherently pleasurable (Lee, 1977). Shared leisure time enhances marital satisfaction as couples usually pursue activities that they both enjoy (Kingston & Nock, 1987). As couples may often have similar interests and hobbies, the interactions during their joint leisure time are likely to exhibit high leisure compatibility between the two individuals (Houts, Robins, & Huston, 1996).

The enhancement of a couple's marital satisfaction through leisure involves several elements. While Lee (1977) suggested that the quantity of time spent together contributes to marital satisfaction, others also argued that the level of interaction is important as well (Holman & Jacquart, 1988). Joint leisure activities with high interaction level have been found to be the most effective in enhancing marital satisfaction (Orthner, 1975). This type of leisure activity tends to be communal and prompts couples to communicate more (Crawford, Houts, Huston, & George, 2002). Further, a positive relationship between leisure and marital satisfaction is developed when the couple mutually like the activity that they engage in (Crawford et al., 2002).

In addition, spousal support also plays a vital role, as respecting each other's leisure interests results in strengthened marital satisfaction. However, it is also possible that spending leisure time together, especially during extended holiday travel, may instigate conflicts between married couples. Barlés-Arizón, Fraj-Andrés, and Martínez-Salinas (2013) suggest that men and women are interested in different activities during travel, with the former looking for sports and action-oriented activities and the latter for cultural and educational experiences. Therefore, when a spouse fails to recognize or accept the partner's leisure interests, his or her disengagement from that activity leads to conflict, which in turn negatively affects their marital satisfaction (Goff, Fick, & Oppliger, 1997).

Notwithstanding the requirement of being away from one's usual home environment, vacation travel as one form of leisure contributes to enhanced marital satisfaction. Gilbert and Abdullah (2004) found that couples who took annual holidays together showed a relatively increased sense of subjective wellbeing and enhanced happiness than those who did not. Vacation travel allows couples to be more intimate than when they are at home following the daily routine (U.S. Travel Association, 2015). The intimacy starts with joint decision-making, as proposed by Nichols and Snepenger (1988). Their study suggested that when every family member took part in the decision-making process, most of them got to enjoy activities they liked. As such, travel decision-making is not a 'one-directional linear progression' (Song, Sparks, & Wang, 2017, p. 258) and can be shared between couples.

Joint decision-making processes may influence marital satisfaction as couples attempt to turn disagreement into consent. Craddock (1991) noted that the process of resolving conflicts and reaching agreement could enhance marital satisfaction. Culture is a crucial factor in how agreements are achieved in various scenarios. It influences the strategies in decision-making and the role each party plays in the process. For example, the work of Song et al. (2017) on young Chinese tourists suggested that the sense of authority within a group was likely to reduce disagreements; and harmony, a foundational element in Chinese culture and society, affects how young people resolve differences. While spending time in leisure has been found to increase marital satisfaction, the understanding of whether and why vacation travel by married couples could increase marital satisfaction remains elusive.

Marital satisfaction and subjective wellbeing

Early works on marriage have largely dealt with depressive symptoms and marital relationships. The importance of understanding the association between individuals' perceived level of wellbeing and positive marital relationship is the subject of a burgeoning body of literature in recent years (Helms, 2013). Individuals' perceived level of wellbeing is measured by subjective wellbeing, which denotes cognitive and emotional life satisfaction (Keyes, Shmotkin, & Ryff, 2002). Existing literature collectively suggests that married individuals exhibit greater subjective wellbeing than never-married individuals or previously married individuals (Gove, Style, & Hughes, 1990). However, it is not the marital status *per se* that enhances subjective wellbeing. It is the satisfied marriage that contributes to increased sense of wellbeing (Proulx et al., 2007).

There are specific benefits that are derived from a satisfied marriage and contribute to an enhanced sense of wellbeing (Mastekaasa, 1994). First, the roles that marriage offers, such as that of the husband or wife, enhance individuals' self-esteem (Crosby, 1987). Self-esteem enables individuals to better manage stress and other negative energies on their mental health, as it provides rewards, status, and avenues to enrichment (Gore & Colten, 1991). Successfully fulfilling the spousal role provides individuals with a sense of identity and self-worth (Gove & Umberson, 1985).

Second, equality in a marital relationship provides emotional support for couples (Sternberg & Grajek, 1984). Equality is established when partners both give and receive benefits (Van Yperen & Buunk, 1990). The emotional support gained in a marital relationship, such as the sense of being cared about and valued as a person, tends to reduce the chance of mental illness (Brown, Nesse, Vinokur, & Smith, 2003). Third, marital satisfaction leads to interpersonal closeness that eventually results in happiness (Laurenceau, Barrett, & Rovine, 2005). While literature on the relationships between marital satisfaction and subjective wellbeing in general is abundant, the current study contextualizes the relationships away from a couple's usual home environment to include the examination of the role of vacation travel.

Chinese marital satisfaction and subjective wellbeing

An increasing number of Chinese are embracing some marriage values considered important by Westerners, such as love and mutual respect (Xia & Zhou, 2003), while their traditional cultural values are still ingrained in their marital dynamics. As with Westerners, Chinese married couples are more likely to experience happiness and accomplishment than divorced, separated, or widowed individuals (Shu & Zhu, 2009). Moreover, marital distress is detrimental to a couple's personal health in the West, as well as in Asian countries such as China (Miller et al., 2013). While these findings signal that the relationship between marital satisfaction and wellbeing in China may be similar to that of Westerners, there is a need to examine the unique characteristics of Chinese marriage, as Chinese values and traditions still guide modern-day life.

China is traditionally a patriarchal society, in which gender roles defined by society influence the dynamics between husband and wife (Yan, Wu, Ho, & Pearson, 2011). The perception of intimacy and sexuality between married couples may be different from that in Western societies due to the conservative nature of the Chinese culture (Liu, Dennis, & Edwards, 2015). Some existing studies have reported that there is a gap between China and Western countries in what married couples consider to be an essential indicator of marital satisfaction. For instance, Pimentel (2000) found that while Chinese couples did not consider spending free time with each other as important for their marital satisfaction, Westerners viewed otherwise.

According to the dyadic adjustment scale (Busby, Christensen, Crane, & Larson, 1995) developed for couples in the United States, one dimension to measure marital satisfaction is the frequency of kissing. In China, however, it is difficult to measure kissing because it may not be as commonly displayed in public as in the West, although hugging in public is becoming acceptable (Shek & Cheung, 2008). Thus, it is imperative to investigate the drivers for Chinese marital satisfaction that reflect current trends under China's gradual Westernization and its persisting traditional Chinese values that jointly influence the marital dynamics of Chinese couples. To date, there are no studies that examine marital satisfaction for the Chinese in the context of vacation travel. The current study investigates factors that enhance couples' marital satisfaction and their subjective wellbeing as the result of vacation travel, as well as the linkages among these three constructs for a Chinese sample.

Methodology

The study drew a sample of 22 Chinese respondents for personal interviews in a Midwestern university community in the United States during January and February 2016. A convenience snowball sampling method was adopted to recruit married Chinese who had traveled with their spouse, with or without other travel companions. The convenience sampling mechanism is appropriate when constraints hinder researchers from recruiting participants in a standard sampling approach (Heckathorn, 2011).

One notable character of this study was the temporary nature of the interviewees' residency in the United States, which suggests their familiarity with and openness to values and beliefs from both the East and West. This was particularly relevant to and informative about the representative population of the study, as such openness might lead to one's views on marriage and subjective wellbeing that reflect cultural traits from both the Eastern and Western value systems. Following the convention of convenience snowball sampling, the first set of interviewees were recruited from the investigators' personal and professional networks and these interviewees were asked to recommend other participants.

Most of the interviewees were female (61%). Half of them were in their 30s, 32% their 20s, and 18% their 40s. Each interviewee had at least a bachelor's degree and a good command of English. Table 1 outlines the sample's demographic profile. Prior to the interviews, which were conducted entirely in English, the interviewees were approached to ensure they were comfortable with the interview format and types of questions. The length of the interviews varied from 30 to 60 minutes. The interviewees were asked questions related to their travel experiences, marital relationship, and marital satisfaction before and after their vacation travels. The interview sessions were recorded with explicit consent from the interviewees. The recordings were transcribed into the textual data, which was analyzed concurrently with data collection until theoretical saturation was reached.

Thematic analysis was conducted to examine the data following the transcription of the recorded interviews. This is an approach used 'across a range of theoretical frameworks and research paradigms' (Clarke & Braun, 2017, p. 297), and it fitted the objective and data of the current study. Major themes that were indicative of the interviewees' views on the interplay among vacation travel, marital satisfaction, and

Table 1. Demographic profile of interviewees.

Interviewee	Gender	Age	Highest level of education	Duration of marriage (years)	Family status (# of children)
1	Female	36	Bachelor's	5	One
2	Female	38	Master's	9	One
3	Male	43	Doctorate	12	One
4	Female	30	Bachelor's	1	No
5	Male	40	Master's	10	Two
6	Female	30	Master's	3	One
7	Female	29	Master's	1	One
8	Female	37	Bachelor's	9	One
9	Male	29	Bachelor's	2	No
10	Female	28	Bachelor's	7	One
11	Male	38	Doctorate	6	One
12	Female	27	Bachelor's	2	One
13	Female	34	Bachelor's	6	One
14	Male	39	Doctorate	7	One
15	Female	29	Bachelor's	4	One
16	Female	28	Master's	3	One
17	Male	29	Master's	2	No
18	Male	43	Doctorate	11	One
19	Female	36	Bachelor's	9	One
20	Male	28	Bachelor's	1	No
21	Female	33	Bachelor's	5	One
22	Female	40	Doctorate	12	One

subjective wellbeing were identified. The analysis process involved four steps of coding and theme-generation. During the first step, the transcription was read thoroughly by the investigators to ensure clarity and accurateness of the information obtained from the interviews. The process was repeated to enhance the investigators' familiarity with the data. During the second step, the investigators proceeded to initial coding and generated broad themes about the drivers of marital satisfaction and subjective well-being from vacation travel.

The third step involved iterative coding based on the broad themes from the second step. Sub-themes emerged as the transcription was carefully revisited separately by different investigators. Two similar sets of themes were derived and compared to ensure comprehensive understanding of the data. Action verbs, descriptive words, and tones of language were simultaneously considered in the analysis. For instance, in analyzing the account of interviewee #18 on his explanation of how he went to a museum because of his wife who loved arts, the investigators also considered his effort made to accommodate his wife and his sense of happiness and excitement implied from his narrative. During the fourth step, underlying dimensions within the sub-themes surfaced as further analysis was employed. The results were collectively examined by the investigators to reach a consensus.

Decrop (2004) suggests four criteria to ensure the trustworthiness of qualitative research. They are credibility, transferability, dependability, and confirmability. The current study adopted four corresponding processes to meet the criteria. First, the study recognized that a well-established process was essential to ensure credibility (Shenton, 2004). So the interview data were analyzed strictly in accordance with the steps required of thematic analysis as described in the previous section. Second, although generalization is not desired in qualitative research, Decrop (2004) argues that the capability for such research to pass on theoretical propositions to other contexts is reasonable. Thus, this study documented in detail the settings of the research and systematically described how the findings were similar to, or different from, existing literature. The third and fourth criteria of dependability and confirmability require the match between what is understood from data, what actually happened, and what is to be accomplished through data analysis. These two criteria were addressed by following corresponding guidelines of Decrop (2004). Initially, the interviews were recorded and the transcriptions were coded collectively by two investigators. Then the coding process and findings were audited and verified by the third to improve the overall dependability and confirmability.

Results

Eleven themes were identified as the benefits from vacation travel that led to enhanced marital satisfaction. These themes were further aggregated into four drivers. They are (a) *Mutual Devotion*, couples becoming more committed to each other and considerate of their spouse's needs and wants; (b) *Reignited Passion*, restoring their love; (c) *Strengthened Bond*, developing deeper intimacy with their spouses; and (d) *Open Communication*, communicating with their spouse better than before trip.

Drivers of enhanced marital satisfaction

Mutual devotion

Individuals' relationships with their spouses are strengthened as they become more devoted to each other during the trip than in their daily life. While keeping relationships stable requires substantial efforts, couples may fail to give full attention to each other with everyday stress and distractions. In this case, vacation travel appears to be the opportunity to compensate for the lack of attention in their daily life. The driver of *Mutual Devotion* encompasses three benefits, including *being considerate of one's partner, empathizing with one's spouse*, and *equality* as the result of vacation travel.

The majority of the interviewees reported that the primary reason that they tried to be considerate of their spouse was to ensure a good trip. That is, the interviewees were more respectful of their spouse's thoughts and behaviors, and cared more about them during vacation travel. Interviewees acknowledged that they remained understanding to avoid ruining their trips due to high trip costs. A 36-year-old female (interviewee #19) mentioned, 'I am usually in a good mood while traveling, so it is natural to treat my spouse well.' It may seem superficial to cite external factors such as trip costs to show caring for their partners. Yet, initiating a dedicated mood is helpful to enhance the marital relationship. Regardless of the triggers, what matters the most is being a good partner. Giving and receiving kindness are equally effective in generating happiness in marriage.

The interviewees also reported that it was highly likely for them to empathize and sympathize with their spouses during the trips. They acknowledged and accepted one another's interests and enjoyed them together. The testimony of a 43-year-old male (interviewee #18) attests to how interviewees tried to sympathize with their spouses while traveling, which leads to an enhanced relationship:

> Being together just makes me happy. My wife majored in Arts, so she loves art, but I have no interest in it. But still, I do go to museums because my wife likes it. When we went to museums together, I could sympathize with her as we talk about what was good, what was most memorable.... and she seemed to be happy. When she likes the trip, I feel happier.

When one's spouse showed empathy, he or she was satisfied because the spouse was willing to participate in the activity of his or her interest, as a 30-year-old female (interviewee #6) noted. They were appreciative of each other's effort and the time they spent together for the activity. This supports the argument from previous research that emotional intelligence—how much couples emotionally understand each other—influences marital satisfaction (Dehkordi, 2012).

Many female interviewees felt that they felt to be an equal partner with their spouse while on vacation travel. As in the case of a 27-year-old female interviewee (#12), she mentioned how her husband would do many things for her, from trip planning to heavy packing tasks. This corresponds to testimonies from the male interviewees who admitted that they tried to do more work than their wife on their trips. For instance, a 29-year-old male (interviewee #17) discussed how he would volunteer in taking more responsibilities as he felt that he needed to make up for being unable to help his wife enough with housework. As a result, women felt that they were treated more equally, while men took pride in themselves being a good husband. The equality and pride together led to a better relationship, which is reflected in the account by a 30-year-old female (interviewee #4):

> Once, we arrived in the hotel very late. However, we were assigned to an uncleaned room. We were both very frustrated because we were very tired after our long flight. But he told me to just relax, and that he would take care of everything and told me to not worry. I was able to just relax, and he did take care of everything.

Reignited passion

Another keystone to an enhanced marital satisfaction is represented by the driver of *Reignited Passion*. Two vacation travel benefits comprise this driver: *reminder of one's affection* and *appreciation*. Being grateful in having each other made couples realize how much they loved each other. As shown in the following account of a 40-year-old male (interviewee #5), specific cues, such as romantic mood during travel, allowed people to realize their affection for their spouse, eventually leading them to resume the expression of love that was forgotten, such as holding hands.

> When we travel as a family, we like to travel in nature. Once, when our family made a trip to this beautiful park, I was reminded of one of the dates I had with my wife. I remembered how much I loved her by then, and much more now. Walking together we got to naturally hold our hands, just like we used to do before we had our kids. We walked, and the kids walked by themselves, so we had our own time.

The interviewees reported growing fonder of their spouses during vacation travel as they were given an opportunity to appreciate having each other. In a stressful daily life, one tends to take his or her spouse for granted. During travel, they realize how much their spouse means to them. 'Simply thinking I am happy to be traveling with my spouse,' as explained by a 29-year-old male (interviewee #9), induced him to appreciate having her by his side. This can be explained by the spillover effect of 'leisure mood enhancement' on couples' relationships (Iwasaki & Mannell, 2000, p. 168). Leisure time helps individuals to be free from everyday stress, which in turn enhances positive mood and reduces negative mood. Being in a good mood helps couples to be more engaged in a romantic relationship, with no interruptions such as stress.

Strengthened bond

Through vacation travel, married couples strengthened their marital bond that positively affected their marital relationship and satisfaction. This driver is composed of three benefits, including *mutual happiness, intimacy*, and *sense of togetherness* as the result of vacation travel. These sentiments developed during a trip were passed on to their attitude towards their spouse after returning home. One of the most salient benefits from vacation travel that strengthened their bond with a spouse was the sense of *mutual happiness*, in which they shared the feeling of happiness with each other. As couples didn't have enough time to spend with their spouse in the daily setting, several interviewees admitted that just being together with their spouse made them happy. Some of the interviewees explained that they enjoyed seeing their spouse being happy while traveling. Moreover, explicitly expressing their happiness made the couple realize how much they cared for each other. Being happy together, not just for themselves but for each other, further strengthened their attachment to each other. This is exemplified in the following account of a 33-year-old female (interviewee #21):

> We were in Thailand last summer. We enjoyed the food, nice hotel, relaxing environment ... but the best moment of the trip for me was seeing how he was enjoying his vacation. He couldn't be happier. I hadn't seen him smiling like that for years. At home, he was always stressed and tired from work. We had no worries while we were on vacation. I was really happy to see him relaxed.

The benefit of *intimacy* developed during vacation travel also contributed to building a stronger bond between partners as individuals became fonder of their spouse. In other words, interviewees felt being emotionally and psychologically closer to their spouse than before. This sentiment was salient for female interviewees who felt secure as their husband took care of the problems that arose during travel. It resulted in greater attachment to their spouse. *Sense of togetherness* helped married couples to form a greater bond as well. Regardless of what they did while traveling, the joint activities *per se* made their experience special. Having only themselves in an unfamiliar setting also triggered this sentiment. Moreover, *sense of togetherness* became prominent when interviewees encountered issues over which that they needed to cooperate and be supportive of each other. The accomplishment of doing something together enhanced their marital satisfaction.

Open communication

This is the fourth driver that enhanced couples' marital satisfaction. It consists of three benefits, including *increased frequency of talk, common topics to talk*, and *constructive topics* from vacation travel. Communication is the first step in meeting spouse's marital expectations, which has been known to contribute to marital satisfaction (Fletcher et al.,

2000). As couples actively and explicitly express their thoughts and feelings, they get to understand each other better. Most of the interviewees agreed that their frequency of communication with their spouse increased while traveling. As the couple spent their days together, they tended to have more conversation than at home. Couples were also forced to actively communicate during the pre-trip stage as they planned for specific travel activities. During their post-trip reminiscence, they had common topics to talk about. Dialogues were often related to their happy moments during travel, which in turn strengthened their bond. As such, sharing similarity with one's spouse (e.g., sharing values, time together, or information) impacted individuals' marital satisfaction, as previously reported in non-travel settings (Chi et al., 2013).

Interviewees also noted that, during vacation travel, they talked about different topics from their daily life. At home, most of their conversations focused on parenting and maintaining the order of the household. However, travel granted them an opportunity to discuss deeper and personal subjects. Novel contextual cues of being in somewhere new and different triggered interviewees to talk about something different. These dialogues allowed them to express their feelings more explicitly, which helped them to understand each other better, as observed in the account of a 40-year-old female (interviewee #22):

> When I travel, I think I become more expressive. I tell my husband what I want to do, what I want to eat. ... In the daily life, I am busy to take care of my kid and my husband. But the travel time for just the two of us makes me to be like ... in some way like a baby. I expect my husband to do what I demand for. At the same time, we get to talk more about our future. Our plans for us, even what we should do after we retire. Our daily conversations when we are at home are very much focused on the problems in raising my son, financial issues ... I don't feel like bringing up those issues while traveling. We only talk about our rosy future.

Outcomes of improved subjective wellbeing

Three outcomes of the subjective wellbeing of married couples as the result of vacation travel emerged from the data. They are *Social Support*, having one's spouse in times of physiological and emotions needs; *Self-esteem*, the evaluation of one's worthiness; and *Happiness*, a feeling of satisfaction. Each outcome encompasses the underlying benefits that explain how vacation travel may improve married couples' subjective wellbeing. The outcome of *Social Support* speaks to individuals' sense of being emotionally cared for and unconditionally supported by their spouse, as well as of offering such support back to their spouse. This outcome comes from three benefits from vacation travel. They are *empathizing with one's spouse, sense of togetherness*, and *equality*. As interviewees' spouses empathized with them during their vacation travel, they were assured that they had someone by their side who supported them. This was attested by a 38-year-old female (interviewee #15):

> When traveling with my husband, I feel secured that somebody is with me. I am not alone. I feel like we are part of a team. We are the co-leaders and we always have each other.

At the same time, interviewees were happy to be the provider of emotional support for their spouse during their trip. In line with this finding, previous studies argued that being emotionally supportive for one's spouse is strongly related to physical health. For instance, Brown et al. (2003) found that providing emotional support to one's spouse reduced mortality rate. *Sense of togetherness* developed from vacation travel also made individuals feel that they were part of their own social circle. That is, their innate needs of belonging (i.e., the desire to be attached to a social group) is fulfilled (Baumeister & Leary, 1995). Interviewees noted that they felt strongly satisfied as they both gave and received support. For instance, a 34-year-old female (interviewee #13) admitted that while she did more housework at home, her husband did most of the work demanded during traveling, which made both of them feel in a relationship with balanced power.

Defined as individuals' overall evaluation of their worth and competence (Weiten, 2004), *Self-esteem* was found to be the second outcome of improved subjective wellbeing from vacation travel. According to the interviewees' account, *Self-esteem* was improved as individuals became more considerate of their partner while traveling. They took care of their spouse better on the trip than when at home. In turn, they took pride in being able to be a better husband or wife. Such improved *Self-esteem* especially applied to men, who took pleasure in managing the overall trip for their spouse. As men tried to be more caring for their wives for certain tasks, such as trip planning and arrangement, they felt a great sense of achievement and self-worth. As a 38-year-old male (interviewee #11) explained, men highly regarded themselves in being the good husband. This sense of accomplishment therefore helped them build a positive sense of self-regard. The finding confirms the existing notion that men's subjective wellbeing is enhanced through marriage as they attain role-related benefits (Greenberger & O'Neil, 1993).

Mutual happiness and *intimacy* were the benefits from vacation travel that formed the driver of *Strengthened Bond* in enhanced marital satisfaction. They were also found to underlie the third outcome of improved subjective wellbeing, *Happiness*. Interviewees reported being happy during their vacation travel because they felt closer with their spouse through shared activities. Accordingly, Interviewee #8 explained how happy she was when she was kayaking with her husband. They didn't share common hobbies at home. Yet, kayaking, paddling together, and looking for directions increased their sense of *intimacy* as well as *happiness*. This is aligned with findings from existing literature that personal happiness is most highly influenced by social relationship with one's intimate partner (Fowler & Christakis, 2008). Specifically, the quality of the relationship, such as how close they are to each other, shapes their level of happiness (Reis & Gable, 2003). Furthermore, sharing happiness with one's spouse also enhances individuals' subjective wellbeing. *Mutual happiness* triggers individuals to be more explicit in expressing their delight, which maximizes their joy.

The duality of benefits from vacation travel

The findings on the four drivers of enhanced marital satisfaction and the three outcomes of improved subjective wellbeing are summarized in Table 2. Their relationships are illustrated in Figure 1. The larger circle contains 'benefits for marital satisfaction from vacation travel,' which together form the four drivers of enhanced marital satisfaction. The smaller circle within the larger one encompasses 'benefits for subjective wellbeing from vacation travel,' which underlie the three outcomes of improved subjective wellbeing. The items in the smaller circle contribute to both.

Table 2. Summary of drivers, outcomes, and vacation benefits.

	Drivers/outcomes	Vacation benefits
Enhanced Marital Satisfaction (Drivers)	Mutual Devotion	being considerate of one's partner
		empathizing with one's spouse
		equality
	Reignited Passion	reminder of one's affection
		appreciation
	Strengthened Bond	mutual happiness
		intimacy
		sense of togetherness
	Open Communication	increased frequency of talk
		common topics
		constructive topics
Improved Subjective Wellbeing (Outcomes)	Social Support	empathizing with one's spouse
		sense of togetherness
		equality
	Self-esteem	being considerate of one's spouse
	Happiness	intimacy
		mutual happiness

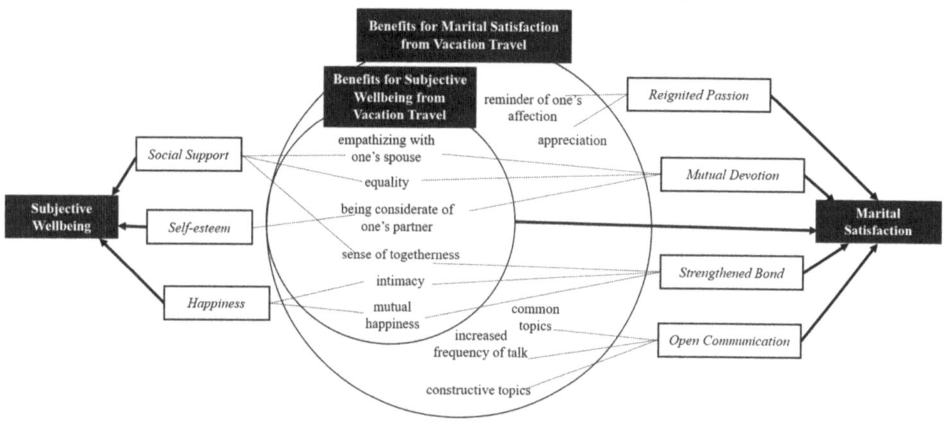

Figure 1. Linkages between vacation travel, marital satisfaction, and subjective wellbeing.

Benefits that are unique to the four drivers of enhanced marital satisfaction include *reminder of one's affection, appreciation, common topics, increased frequency of talk*, and *constructive topics.* Meanwhile, every benefit that contributes to one of the three outcomes of improved subjective wellbeing also accounts for one of the four drivers of enhanced marital satisfaction. These overlapping benefits are *empathizing with one's spouse, equality, being considerate of one's partner, sense of togetherness, intimacy*, and *mutual happiness.* Enhanced marital satisfaction and improved subjective wellbeing from vacation travel highlight the importance of quality time between couples. As Berg, Trost, Schneider, and Allison (2001) noted, high-quality time spent together between couples was essential for a happy marriage, which is reflected in marital satisfaction and subjective wellbeing in the current study. Traveling with one's spouse for a vacation is an effort made to spend quality time together. It does not seem to matter what specific activities they are jointly engaged in.

The Chinese couples interviewed for the study were seeking benefits for themselves, as well as for their spouses, towards the goal of enhanced marital satisfaction. For instance, while individuals sought empathy from their partner, they also took pleasure in empathizing with their spouse. The mutual gratification allowed the husband and wife to emotionally support and complement each other. This confirms the notion that the primary aim in marriage is a communal relationship that meets each other's needs (Clark & Mills, 1994). Providing support to one's partner is as important as receiving support from a partner in enhancing subjective wellbeing (Davis, Morris, & Kraus, 1998).

The linkages among vacation travel, marital satisfaction, and subjective wellbeing were found from the interviewees' liminal experiences that vacation travel offered. As tourists are placed in a liminal space, they are freed from the distractions from the daily life and are more relaxed (Selanniemi, 2003). Such a diversion allows the couples to focus more on each other. Therefore, they become more caring and devoted towards their spouse, which translates into enhanced marital satisfaction as well as improved subjective wellbeing. As Cohen (1974) argues, liminal experiences that take place outside of the everyday space and circumstance encourage individuals to be more expressive when engaging in non-ordinary activities. In the case of vacation travel of married couples, being more expressive to each other facilitates open communication, which was found to be both a driver for enhanced marital satisfaction and an outcome of improved subjective wellbeing.

Conclusion

In today's Chinese society, the spousal relationship has become one of the most important determinants for a strong marriage. However, various issues exist affecting the bond between couples. Conflicts have been found to be prevalent among young Chinese couples who were born in the 1980s and 1990s and grew up as a single child in their families. In particular, Rosenberg and Jing (1996) suggested that China's one-child policy had created spoiled children, who tend to prioritize their own interests instead of others'. This may have a significant impact on how young couples manage their marriages. Traveling together allows couples to re-establish the bond that might have been eroded in daily married life. In this light, the current study has identified the drivers for enhanced marital satisfaction and outcomes of improved subjective wellbeing in the context of vacation travel. These drivers are *Mutual Devotion, Reignited Passion, Strengthened Bond,* and *Open Communication*; and the outcomes are *Social Support, Self-esteem,* and *Happiness.* As such, the findings of this study bear both theoretical and practical implications.

The current research expands the understanding of the role of vacation travel in marital relationships of Chinese married couples. While connections between family relationships and vacation travel have been extensively examined, linkages between vacation travel and the spousal relationship have been relatively understudied. The study reveals the unique benefits of vacation travel on enhancing marital satisfaction and subjective wellbeing. For instance, it suggests that married individuals tend to make decisions in favor of their spouses on vacation trips. They are likely to show dedication to each other, which leads to enhanced marital satisfaction.

Mutual Devotion results in positive subjective wellbeing, through which *Social Support* is established among individuals. For instance, both marital satisfaction and subjective wellbeing is improved when couples empathize with their spouse. The results of this study advance knowledge on the outcomes of travel. While previous studies discussed various outcomes of travel, such as augmented wellbeing from vacation travel (Gilbert & Abdullah, 2004), the current study reveals an increase in marital satisfaction as an additional benefit of travel. As tourists today not only travel to entertain themselves, but also seek meaningful experiences from their trips (Pearce & Packer, 2013), the study's affirmation that individuals' marital satisfaction could be enhanced from vacation travel is a significant contribution to the literature on family, travel, and wellbeing.

Marriage in contemporary China has been undergoing transformations due to changing socio-economic conditions and the convergence of Chinese traditions and Western values. The study's finding that vacation travel enhances marital satisfaction for Chinese couples validates prior reports on other population groups (Durko & Petrick, 2016). For instance, marital satisfaction tended to be high for Western couples who were engaged in frequent and high quality conversation (De Bloom, Geurts, & Kompier, 2013). This was true for the Chinese sample in this study as well. The role of communication for Chinese couples seems to be a product of Western influence. It emphasizes the importance of love rather than a convenient marriage pre-arranged by families. While Chinese couples exhibit similarities to their Western counterparts, their traditional values remain dominant. In both rural and urban areas of modern-day China, men are likely to be the decision-maker and the gender gap persists (Xu, Li, & Yu, 2014).

However, the study's findings suggest that this discrepancy appear to be shrinking when Chinese couples are on vacation and both the husband and wife participate in decision-making. The gender role in decision-making changes when it comes to vacation travel, which may serve as a gateway for females to express and implement their own opinions. That the participation of Chinese wives in vacation decision-making has become more prevalent may be a differentiating perspective between China and the West about how specifically vacation travel enhances marital satisfaction. While the change in gender roles in vacation decision-making is one factor that increases marital satisfaction for Chinese couples, it has been rarely reported in previous studies regarding Western respondents.

The results on the role of vacation travel in marital satisfaction and subjective wellbeing present practical implications for policy-makers, the travel trade, and other service providers in the tourism system. Miller and Fang (2012) suggest that China's economic reforms since the late 1970s have led to drastic changes in Chinese daily life, affecting the 'structure and attitudes of the Chinese family' (p. 174). At the same time, the one-child policy has created both economic and social pressure for parents and grandparents in child education and upbringing in general. This has caused a rise in demand for mental health counseling services in China, especially for married couples (Chang, Tong, Shi, & Zeng, 2005) The rising divorce and remarriage rates as observed by Wang and Zhou (2010) signal a phenomenon that is in fact against the 'Confucian ideal of harmony' (Naustdalslid, 2014, p. 305).

The findings from the research should be informative in a discourse on how to incorporate vacation travel into policy-making in China pertaining to social issues, such as the relationship between quality of life, marital satisfaction, and subjective wellbeing. In general, policy-makers may promote vacation travel for married couples to fortify their relationships. In devising strategies to encourage romantic and family travel, studies should be encouraged to investigate what hinders married couples from traveling together and how to accommodate their needs. In addition, understanding the factors that change marital satisfaction and subjective wellbeing from vacation travel should enable travel trade and destination organizations to better design and deliver their goods and services for romantic and family travelers.

The conclusions from this study, and implications thereof, should be interpreted in view of the limited number of interviewees in the sample, albeit with the sample's unique profile and forward-looking attempt to represent Chinese society. The interviewees do not represent a wide spectrum of demographic characteristics. In addition, because they were recruited in a setting of a university community in the United States, their education level does not represent that of the Chinese population. Also, Chinese marriages are diverse and formed by regional cultures and customs (Ji & Yeung, 2014). The findings from this study are therefore not generalizable to interpret the interplays of marital satisfaction, vacation travel, and subjective wellbeing across different geographical locations in China. This study explores such perspectives with only one segment of the population represented by the sample. In addition to regional differences, future studies may examine the long-term effects of vacation travel on sustaining the quality of marital relationships.

Another direction for future research is to investigate how vacation travel facilitates marital satisfaction and subjective wellbeing for couples in 'late marriages,' a Chinese concept that describes couples that get married after their mid-20s or early 30s. To (2015) suggests that professionals, particularly females, are not attracted to the idea of traditional marriage due to their economic independence. It may imply substantial differences in their view of marital satisfaction after they get married compared with those who do get married 'on time'. Consequently, whether vacation travel is beneficial to marital satisfaction and subjective wellbeing remains a question for this population as well.

It is also necessary for future studies to examine how conflicts that arise during vacation travel influence one's marital satisfaction and subjective wellbeing. None of the interviewees in this study revealed or discussed any disagreements and conflicts throughout their trips. This could be due to the fact that data were mostly collected via face-to-face interviews. Although the interviewees were approached in a friendly manner, were ensured confidentiality, and talked to with accommodative tones, it might still be difficult for them to disclose sensitive and personal information on their domestic lives. Future studies may incorporate an anonymous form of interview using phone calls or online surveys.

The economic impacts of Chinese tourists on the global market have gained much attention. Yet few studies have examined how vacation travel has changed the lives of the Chinese tourists themselves. Because family acts as the basis of the Chinese society, the current study and its findings implicate a necessary link between travel, family, and social stability in China. It is in the interest of policy-makers and social scientists to explore how vacation travel may inspire social changes in China in relation to gender roles and family structure, as well as how the wellbeing of Chinese society as a whole is affected by such changes.

Disclosure statement

No potential conflict of interest was reported by the authors.

References

Abelmann, N., & Kang, J. (2014). A fraught exchange? US media on Chinese international undergraduates and the American university. *Journal of Studies in International Education, 18* (4), 382–397.

Addis, J., & Bernard, M. E. (2002). Marital adjustment and irrational beliefs. *Journal of Rational-Emotive & Cognitive-Behavior Therapy, 20*(1), 3–13.

Aghajanian, A., & Thompson, V. (2013). Recent divorce trend in Iran. *Journal of Divorce & Remarriage, 54*(2), 112–125.

Amato, P. R. (2000). The consequences of divorce for adults and children. *Journal of Marriage and Family, 62*(4), 1269–1287.

Aref, F. (2011). The effects of tourism on quality of life: A case study of Shiraz, Iran. *Life Science Journal, 8*(2), 26–30.

Barlés-Arizón, M., Fraj-Andrés, E., & Martínez-Salinas, E. (2013). Family vacation decision making: The role of woman. *Journal of Travel & Tourism Marketing, 30*(8), 873–890.

Bar-On, R., Handley, R., & Fund, S. (2006). The impact of emotional intelligence on performance. In V. Druskat, F. Sala, & G. Mount (Eds.), *Linking emotional intelligence and performance at work* (pp. 3–19). Mahwah, NJ: Lawrence Erlbaum Associates, Inc.

Batool, S. S., & Khalid, R. (2012). Emotional intelligence: A predictor of marital quality in Pakistani couples. *Pakistan Journal of Psychological Research, 27*(1), 65–88.

Baumeister, R. F., & Leary, M. R. (1995). The need to belong: Desire for interpersonal attachments as a fundamental human motivation. *Psychological Bulletin, 117*(3), 497–529.

Beach, S. R., Sandeen, E., & O'Leary, K. D. (1990). *Depression in marriage: A model for etiology and treatment.* New York, NY: Guilford Press.

Berg, E. C., Trost, M., Schneider, I. E., & Allison, M. T. (2001). Dyadic exploration of the relationship of leisure satisfaction, leisure time, and gender to relationship satisfaction. *Leisure Sciences, 23*(1), 35–46.

Bradbury, T. N., Fincham, F. D., & Beach, S. R. (2000). Research on the nature and determinants of marital satisfaction: A decade in review. *Journal of Marriage and Family, 62*(4), 964–980.

Brown, S. L., Nesse, R. M., Vinokur, A. D., & Smith, D. M. (2003). Providing social support may be more beneficial than receiving it results from a prospective study of mortality. *Psychological Science, 14*(4), 320–327.

Busby, D., Christensen, C., Crane, R., & Larson, J. (1995). A revision of the dyadic adjustment scale for use with distressed and nondistressed couples: Construct hierarchy and multidimensional scales. *Journal of Marital and Family Therapy, 21*(3), 289–308.

Campbell, L., Simpson, J. A., Kashy, D. A., & Rholes, W. S. (2001). Attachment orientations, dependence, and behavior in a stressful situation: An application of the actor-partner interdependence model. *Journal of Social and Personal Relationships, 18*(6), 821–843.

Caughlin, J. P. (2002). The demand/withdraw pattern of communication as a predictor of marital satisfaction over time. *Human Communication Research, 28*(1), 49–85.

Chang, D., Tong, H., Shi, Q., & Zeng, Q. (2005). Letting a hundred flowers bloom: Counseling and psychotherapy in the People's Republic of China. *Journal of Mental Health Counseling, 27* (2), 104–116.

Chen, F. M., & Li, T. S. (2007). Marital enqing: An examination of its relationship to spousal contributions, sacrifices, and family stress in Chinese marriages. *The Journal of Social Psychology, 147*(4), 393–412.

Chi, P., Epstein, N. B., Fang, X., Lam, D. O., & Li, X. (2013). Similarity of relationship standards, couple communication patterns, and marital satisfaction among Chinese couples. *Journal of Family Psychology, 27*(5), 806–816.

Cholewinski, R. (2002). Family reunification and conditions placed on family members: Dismantling a fundamental human right. *European Journal of Migration and Law, 4*(3), 271–290.

Clark, M. S., & Mills, J. (1994). Communal and exchange relationships: Controversies and research. In R. Erber & R. Gilmour (Eds.), *Theoretical frameworks for personal relationships* (pp. 29–42). Hillsdale, NJ: Lawrence Erlbaum Associates.

Clarke, V., & Braun, V. (2017). Thematic analysis. *The Journal of Positive Psychology, 12*(3), 297–298.

Cohen, E. (1974). Who is a tourist? A conceptual clarification. *The Sociological Review, 22*(4), 527–555.

Craddock, A. E. (1991). Relationships between attitudinal similarity, couple structure, and couple satisfaction within married and de facto couples. *Australian Journal of Psychology, 43*(1), 11–16.

Crawford, D. W., Houts, R. M., Huston, T. L., & George, L. J. (2002). Compatibility, leisure, and satisfaction in marital relationships. *Journal of Marriage and Family, 64*(2), 433–449.

Crosby, F. J. (1987). *Spouse, parent, worker: On gender and multiple roles.* New Haven: Yale University Press.

Davis, M. H., Morris, M. M., & Kraus, L. A. (1998). Relationship-specific and global perceptions of social support: Associations with well-being and attachment. *Journal of Personality and Social Psychology, 74*(2), 468.

De Bloom, J., Geurts, S. A., & Kompier, M. A. (2013). Vacation (after-) effects on employee health and well-being, and the role of vacation activities, experiences and sleep. *Journal of Happiness Studies, 14*(2), 613–633.

Decrop, A. (2004). Trustworthiness in qualitative tourism research. In J. Phillimore & L. Goodson (Eds.), *Qualitative research in tourism: Ontologies, epistemologies and methodologies* (pp. 156–169). London: Routledge.

Dehkordi, M. (2012). Relationship between emotional intelligence and marital satisfaction in couples. *International Journal of Behavioral Sciences, 6*(2), 161–169.

Durko, A. M., & Petrick, J. F. (2013). Family and relationship benefits of travel experiences: A literature review. *Journal of Travel Research, 52*(6), 720–730.

Durko, A. M., & Petrick, J. F. (2016). Travel as relationship therapy: Examining the effect of vacation satisfaction applied to the investment model. *Journal of Travel Research, 55*(7), 904–918.

Durodoye, B. A. (1997). Factors of marital satisfaction among African American couples and Nigerian male/African American female couples. *Journal of Cross-Cultural Psychology, 28*(1), 71–80.

Fincham, F. D., Beach, S. R., Harold, G. T., & Osborne, L. N. (1997). Marital satisfaction and depression: Different causal relationships for men and women? *Psychological Science, 8*(5), 351–356.

Finucane, M. O., & Horvath, C. W. (2000). Lazy leisure: A qualitative investigation of the relational uses of television in marriage. *Communication Quarterly, 48*(3), 311–321.

Fletcher, G. J., Simpson, J. A., & Thomas, G. (2000). The measurement of perceived relationship quality components: A confirmatory factor analytic approach. *Personality and Social Psychology Bulletin, 26*(3), 340–354.

Fowler, B., & Christakis, N. A. (2008). Dynamic spread of happiness in a large social network: Longitudinal analysis over 20 years in the Framingham heart study. *British Medical Journal, 338*, 1–13.

Gaunt, R. (2006). Couple similarity and marital satisfaction: Are similar spouses happier? *Journal of Personality, 74*(5), 1401–1420.

Gilbert, D., & Abdullah, J. (2004). Holiday taking and the sense of well-being. *Annals of Tourism Research, 31*(1), 103–121.

Goff, S. J., Fick, D. S., & Oppliger, R. A. (1997). The moderating effect of spouse support on the relation between serious leisure and spouses' perceived leisure-family conflict. *Journal of Leisure Research, 29*(1), 47–60.

Gore, S., & Colten, M. E. (1991). Gender, stress, and distress. In J. Eckenrode (Ed.), *The social context of coping* (pp. 139–163). Boston, MA: Springer.

Gove, W. R., Style, C. B., & Hughes, M. (1990). The effect of marriage on the well-being of adults: A theoretical analysis. *Journal of Family Issues, 11*(1), 4–35.

Gove, W. R., & Umberson, D. (1985). *Marriage and the well-being of men and women*. Paper presented at the annual meeting of the American Sociological Association, Washington, DC.

Greenberger, E., & O'Neil, R. (1993). Spouse, parent, worker: Role commitments and role-related experiences in the construction of adults' well-being. *Developmental Psychology, 29*(2), 181–197.

Heckathorn, D. D. (2011). Comment: Snowball versus respondent-driven sampling. *Sociological Methodology, 41*(1), 355–366.

Helms, H. M. (2013). Marital relationships in the twenty-first century. In G. W. Peterson & K. R. Bush (Eds.), *Handbook of Marriage and the Family* (pp. 233–254). New York, NY: Springer.

Holman, T. B., & Jacquart, M. (1988). Leisure-activity patterns and marital satisfaction: A further test. *Journal of Marriage and the Family, 50*(1), 69–77.

Houts, R. M., Robins, E., & Huston, T. L. (1996). Compatibility and the development of premarital relationships. *Journal of Marriage and Family, 58*(1), 7–20.

Iwasaki, Y., & Mannell, R. C. (2000). Hierarchical dimensions of leisure stress coping. *Leisure Sciences, 22*(3), 163–181.

Ji, Y., & Yeung, W. (2014). Heterogeneity in contemporary Chinese marriage. *Journal of Family Issues, 35*(12), 1662–1682.

Johnson, H. A., Zabriskie, R. B., & Hill, B. (2006). The contribution of couple leisure involvement, leisure time, and leisure satisfaction to marital satisfaction. *Marriage & Family Review, 40*(1), 69–91.

Kennedy, S., & Ruggles, S. (2014). Breaking up is hard to count: The rise of divorce in the United States, 1980-2010. *Demography, 51*(2), 587–598.

Keyes, C. L. M., Shmotkin, D., & Ryff, C. D. (2002). Optimizing well-being: The empirical encounter of two traditions. *Journal of Personality and Social Psychology, 82*(6), 1007–1022.

Kingston, P. W., & Nock, S. L. (1987). Time together among dual-earner couples. *American Sociological Review, 52*(3), 391–400.

Klohnen, E. C., & Mendelsohn, G. A. (1998). Partner selection for personality characteristics: A couple-centered approach. *Personality and Social Psychology Bulletin, 24*(3), 268–278.

Laurenceau, J. P., Barrett, L. F., & Rovine, M. J. (2005). The interpersonal process model of intimacy in marriage: A daily-diary and multilevel modeling approach. *Journal of Family Psychology, 19*(2), 314–323.

Lee, G. R. (1977). Age at marriage and marital satisfaction: A multivariate analysis with implications for marital stability. *Journal of Marriage and Family, 39*(3), 493–504.

Lehto, X. Y., Choi, S., Lin, Y.-C., & MacDermid, S. M. (2009). Vacation and family functioning. *Annals of Tourism Research, 36*(3), 459–479.

Leung, K., Koch, P. T., & Lu, L. (2002). A dualistic model of harmony and its implications for conflict management in Asia. *Asia Pacific Journal of Management, 19*(2), 201–220.

Li, T., & Fung, H. H. (2011). The dynamic goal theory of marital satisfaction. *Review of General Psychology, 15*(3), 246–254.

Liu, W., Dennis, J. M., & Edwards, C. P. (2015). Chinese parents' involvement in sexuality education for adolescents. *International Journal of Sexual Health, 27*(4), 485–507.

Mannell, R. C., & Iso-Ahola, S. E. (1987). Psychological nature of leisure and tourism experience. *Annals of Tourism Research, 14*(3), 314–331.

Mastekaasa, A. (1994). Marital status, distress, and well-being: An international comparison. *Journal of Comparative Family Studies, 25*(2), 183–205.

Miller, J. K., & Fang, X. (2012). Marriage and family therapy in the People's Republic of China: Current issues and challenges. *Journal of Family Psychotherapy, 23*(3), 173–183.

Miller, R. B., Mason, T. M., Canlas, J. M., Wang, D., Nelson, D. A., & Hart, C. H. (2013). Marital satisfaction and depressive symptoms in China. *Journal of Family Psychology, 27*(4), 677–682.

Naustdalslid, J. (2014). Circular economy in China – The environmental dimension of the harmonious society. *International Journal of Sustainable Development & World Ecology, 21* (4), 303–313.

Nichols, C. M., & Snepenger, D. J. (1988). Family decision making and tourism behavior and attitudes. *Journal of Travel Research, 26*(4), 2–6.

Oliver, R. L., & Swan, J. E. (1989). Consumer perceptions of interpersonal equity and satisfaction in transactions: A field survey approach. *Journal of Marketing, 53*(2), 21–35.

Orthner, D. K. (1975). Leisure activity patterns and marital satisfaction over the marital career. *Journal of Marriage and Family, 37*(1), 91–102.

Pearce, P. L., & Packer, J. (2013). Minds on the move: New links from psychology to tourism. *Annals of Tourism Research, 40*, 386–411.

Pimentel, E. E. (2000). Just how do I love thee? Marital relations in urban China. *Journal of Marriage and Family, 62*(1), 32–47.

Proulx, C. M., Helms, H. M., & Buehler, C. (2007). Marital quality and personal well-being: A meta-analysis. *Journal of Marriage and Family, 69*(3), 576–593.

Reis, H. T., & Gable, S. L. (2003). Toward a positive psychology of relationships. In C. L. Keyes & J. Haidt (Eds.), *Flourishing: The positive person and the good life* (pp. 129–159). Washington, DC: American Psychological Association.

Rosenberg, B., & Jing, Q. (1996). A revolution in family life: The political and social structural impact of China's one child policy. *Journal of Social Issues, 52*(3), 51–69.

Ryan, C. (1994). Leisure and tourism—The application of leisure concepts to tourist behavior—A proposed model. In A. Seaton (Eds.), *Tourism state of the art* (pp. 294–307). Chichester, England: John Wiley & Sons.

Schwartz, J. (2005). The socio-economic benefits of marriage: A review of recent evidence from the United States. *Economic Affairs, 25*(3), 45–51.

Selanniemi, T. (2003). On holiday in the Liminoid Playground: Place, time, and self in tourism. In T. Bauer & B. McKercher (Eds.), *Sex and tourism: Journeys of romance, love, and lust* (pp. 19–29). New York, NY: Haworth Press.

Sharaievska, I., Kim, J., & Stodolska, M. (2013). Leisure and marital satisfaction in intercultural marriages. *Journal of Leisure Research, 45*(4), 445–465.

Shek, D. T., & Cheung, C. K. (2008). Dimensionality of the Chinese Dyadic Adjustment Scale based on confirmatory factor analyses. *Social Indicators Research, 86*(2), 201–212.

Shenton, A. K. (2004). Strategies for ensuring trustworthiness in qualitative research projects. *Education for Information, 22*(2), 63–75.

Shu, X., & Zhu, Y. (2009). The quality of life in China. *Social Indicators Research, 92*(2), 191–225.

Song, H., Sparks, B. A., & Wang, Y. (2017). Exploring disagreement prevention and resolution in travel decision-making of young Chinese travelers. *Journal of Travel & Tourism Marketing, 34* (2), 257–273.

Stack, S., & Eshleman, J. R. (1998). Marital status and happiness: A 17-nation study. *Journal of Marriage and Family, 60*(2), 527–536.

Sternberg, R. J., & Grajek, S. (1984). The nature of love. *Journal of Personality and Social Psychology, 47*(2), 312–329.

Stimpson, J. P., Wilson, F. A., & Peek, M. K. (2012). Marital status, the economic benefits of marriage, and days of inactivity due to poor health. *International Journal of Population Research*, 1–6.

Stutzer, A., & Frey, B. S. (2006). Does marriage make people happy, or do happy people get married? *The Journal of Socio-Economics*, *35*(2), 326–347.

To, S. (2015). *China's leftover women: Late marriage among professional women and its consequences*. London, UK: Routledge.

U.S. Travel Association. (2015). *Travel can ignite romance and improves intimacy* [Fact sheet]. Retrieved from https://www.ustravel.org/sites/default/files/Media%20Root/5.2015_Relationship_FactSheet.pdf

Van Yperen, N. W., & Buunk, B. P. (1990). A longitudinal study of equity and satisfaction in intimate relationships. *European Journal of Social Psychology*, *20*(4), 287–309.

Voorpostel, M., van der Lippe, T., & Gershuny, J. (2010). Spending time together—Changes over four decades in leisure time spent with a spouse. *Journal of Leisure Research*, *42*(2), 243–265.

Walster, E., Traupmann, J., & Walster, G. W. (1978). Equity and extramarital sexuality. *Archives of Sexual Behavior*, *7*(2), 127–142.

Wang, Q., & Zhou, Q. (2010). China's divorce and remarriage rates: Trends and regional disparities. *Journal of Divorce & Remarriage*, *51*(4), 257–267.

Weiten, W. (2004). *Psychology themes and variations*. Belmont, CA: Wadsworth/Thomson Learning.

Williams, K. (2003). Has the future of marriage arrived? A contemporary examination of gender, marriage, and psychological well-being. *Journal of Health and Social Behavior*, *44*(4), 470–487.

Xia, Y. R., & Zhou, Z. G. (2003). The transition of courtship, mate selection, and marriage in China. In R. Hamon & B. Ingoldsby (Eds.), *Mate selection across cultures* (pp. 231–246). California: Sage.

Xu, Q., Li, J., & Yu, X. (2014). Continuity and change in Chinese marriage and the family. *Chinese Sociological Review*, *47*(1), 30–56.

Xu, X., & Whyte, M. K. (1990). Love matches and arranged marriages: A Chinese replication. *Journal of Marriage and Family*, *52*(3), 709–722.

Yan, E., Wu, A. M., Ho, P., & Pearson, V. (2011). Older Chinese men and women's experiences and understanding of sexuality. *Culture, Health & Sexuality*, *13*(9), 983–999.

Zhou (2017, September 6). Marriage rate down, divorce rate up as more Chinese couples say 'I don't' or 'I won't any more'. *South China Morning Post*. Retrieved from https://www.scmp.com/news/china/society/article/2109868/marriage-rate-down-divorce-rate-more-chinese-couples-say-i-dont

Is Airbnb a Good Choice for Family Travel?

Pearl M. C. Lin

ABSTRACT

This study investigates how a sharing economy shapes family travel in accommodation service. This study draws on a data source composed of a literature review and in-depth interviews with family travelers. Interviews with Airbnb family users are transcribed and analyzed using NVivo. Coding enables the researcher to identify major dimensions of the effects of Airbnb on family travel. The results show that special experience is the most important dimension of family travelers, followed by facilities and space, online reviews, location, and recommendations of friends. This outcome serves as pointers for Airbnb hosts in addressing the needs and expectations in family travel.

AIRBNB 是家庭旅行的好选择吗？

摘要

本研究探讨共享经济如何塑造家庭旅行在住宿服务。本研究借鉴了文献回顾和对家庭旅行者进行深入访谈的数据源。通过 NVivo 对 Airbnb 家庭用户的访谈进行转录和分析。研究人员使用编码确定 Airbnb 对家庭旅行影响的主要方面。结果表明, 特殊经验是家庭旅客最重要的维度, 其次是设施和空间、在线评论、地点和朋友推荐。这一结果是 Airbnb 主人在家庭旅行中满足需求和期望的经营指针。

Introduction

The family travel market is considered a robust growing sector among the various types of tourism business (Park, Lehto, & Park, 2008). Accommodation in family travel is always a challenging task for family decision-makers. In terms of family budget, the choices of accommodation may vary from traditional bed and breakfast, budget hotel, upscale hotel, service apartment, resort, and informal accommodation – namely, Airbnb. The role of decision-makers in tourism may differ according to cultural background, demographics and socioeconomics, and knowledge of product (Kim, Choi, Agrusa, Wang, & Kim, 2010). The stage of family life cycle is closely related with how family travelers manage their vacation expenditure (Lawson, 1991).

Technology development resulted in the transformation of social media into a powerful platform for the hospitality and tourism industries, particularly their online marketing strategies (Lin, Tung, Qiu Zhang, & Gu, 2018). Given the proliferation of social media and the continued growth of transportation services, travelers continuously redefine their modes of travel, the places they stay, and their discovery of new places. The surge of budget airlines and the recent development of the cruise industry enhance the motivation and frequency of travel in Asia and facilitated the emergence of unconventional accommodations and dining styles. The tourism industry provides significant economic benefits. The accommodation sector also plays a key role in overall travel experience.

The sharing economy raises the value and quality of businesses and individuals, affects tourism development, focuses on user needs, and increases employability (Roblek, Stok, & Mesko, 2016). The sharing economy also enables destinations to improve their response to peak demands by offering alternative services (Cheng, 2016). This development results in emerging issues and concerns that challenge tourism stakeholders because of the availability of informal accommodation services, such as Airbnb, in many cities, including Paris, New York, San Francisco, London, Sydney, Amsterdam, and Barcelona. The Airbnb platform influences the decision making of travelers toward room booking and affects stakeholders, including governments, tourism boards, hotels, and residents (Guttentag, 2015). The rapid expansion of Airbnb may also affect the hotel sector. This informal accommodation service has potential negative and positive effects that may directly influence the image of destinations and the direction of future tourism development of cities. Airbnb accommodation can also serve as supplemental hotel inventory in a mega event (Lehr, 2015). For example, over 600,000 foreign fans visited the 2014 FIFA world cup, but only 55,400 hotel beds were available in Rio, Brazil (Panja & Millard, 2014). The Airbnb platform was able to assist with alternative accommodation to support these needs during this event. The sharing economy solves market demand by offering users large amounts of informative data (Nica & Potcovaru, 2015).

Accommodation occupies a major cost in a travel budget. In the past, finding reasonable accommodations depended on hotel sources. For instance, a family of four normally books two rooms, which can be costly. Currently, Airbnb offers various types of accommodations and a great range of prices that family travelers can choose from according to their budget and preference. By shaping the traditional business model, Airbnb creates an online marketplace that features social trust between hosts and guests, wherein users generate comments, identify photos, and create a personal profile using Facebook (Guttentag, 2015). Communication between hosts and guests enables them to share private space via online tools, thereby creating a new form of tourism. This new viral form of marketing through blogs, chatrooms, and portals demonstrates a strong implication to existing service providers (Roblek et al., 2016). This study aims to investigate the key factors of family travelers in using Airbnb accommodation for family travel.

Literature

'Leisure is a primary resource for familial development' (Lehto, Lin, Chen, & Choi, 2012, p. 838). This definition indicates the importance of leisure activities and family cohesion. Family travel is a major segment of the travel and tourism industry (Kang, Hsu, & Wolfe, 2003). Lehto, Fu, Li, and Zhou (2017) used as a basis the findings of four dimensions of vocation benefits from Chinese family travelers; they found that the first level of benefit, namely, experiential learning, pertains to children. The second and third levels of benefit – namely, communication, and togetherness and shared exploration – pertain to family. The fourth level of benefit – namely, escape and relaxation – pertains to self (individual well-being). 'Homely feelings and cultural meanings are fundamental to most family holidays' when finding a suitable accommodation service (Obrador, 2012, p. 409). Parents often sacrifice their wishes and interests for their children; these wishes include destinations, modes, and activities (Decrop, 2005). Nanda, Hu, and Bai (2007) found that families often visit places that are interesting to children. Many studies (Belch, Belch, & Geresino, 1985; Berey & Pollay, 1968; Decrop, 2005; Flurry, 2007; Flurry & Burns, 2005; Khoo-Lattimore, Prayag, & Cheah, 2015; Kozak, 2010; Lehto, Choi, Lin, & MacDermid, 2009; Nanda et al., 2007; Nickerson & Jurowski, 2001; Thornton, Shaw, & Williams, 1997; Wang, Hsieh, Yeh, & Tsai, 2004) have discussed the influential role of different ages of children in stimulating holiday choices. Parents often prioritize the needs of their children. Thus, accommodation is a key element of experience for family travel. Decisions on the place and format of an accommodation service can be influenced by various factors. Hotels and Airbnb offer diverse experiences and services that are closely related to the budget plan of individuals.

Decision-making process in family travel involving young children

Families with children younger than age 11 tend to emphasize family coalition during decision-making to ensure that all members have a good time (Lehto et al., 2012). Their ideal holiday is 'going away from home (push-perspective) than seeing new places and scenery (pull-perspective)' (Gram, 2005, p. 18). Khoo-Lattimore et al. (2015) studied Asian parents with young children with respect to their process of selecting resort hotels and vacation needs. They identified the following essential factors: travel distance, interaction quality, child-friendly amenities, safety, and family-oriented programs. Rothenfluh, Germeni, and Schulz (2016) noted that parents search for hotel location and accessibility, hygiene standards, children-friendly rooms, and sleep quality. Wong, Ap, and Li (2001) discussed the findings of family travel preferences in an accommodation service. They noted that family travelers prefer lodging facilities that provide services and products, such as swimming pools with lifeguards, and discounted packages for families. The demand for family travel in accommodation services has gained attention. Hotel chains started focusing on this market by offering children's programs, activities, and amenities to enrich their overall stay experience (Nickerson & Jurowski, 2001). Choi, Lehto, and Brey (2010, p. 135) found that families with young children 'do not develop loyalty to a particular experience or that as they grow their propensity to develop loyalty increases.' This finding enables practitioners to determine ways on how to increase loyalty given that young children are future spenders in the industry.

Previous studies that examined the process of selecting accommodation services emphasize that cleanliness is the most important factor for travelers from Mainland China and Western countries (Tsai, Yeung, & Yim, 2011). Chinese travelers tend to focus on prices, brand names, and prestigious statuses, whereas Western travelers focus on room size, security, distance, and location (Tsai et al., 2011). Three determining factors are considered by leisure travelers, namely, location, price, and service quality (Tsai et al., 2011). Room rate is often the most crucial factor in the hotel choices of international travelers (Wong & Chi-Yung, 2002). Despite the preferences and cultural differences of travelers in hotel selection criteria, previous studies did not explore the accommodation selection criteria of family travelers in an informal accommodation service, such as Airbnb. In contrast, family travelers are involved in various stages of travel planning, wherein parents may not be the only decision-makers. The service needs of family travelers may differ from those of other groups. Costs may be doubled when traveling with children, based on their age and number. This factor may lead parents to choose alternative accommodation services to meet their budget.

This study argues that alternative accommodation service platforms, such as Airbnb, may match the characteristics and preferences for family travel. What are the key factors that influence family travelers to choose Airbnb accommodations instead of hotels? What kind of experience with Airbnb do they value the most?

Value determinant of Airbnb for family travel

The sharing economy increases consumer welfare and opens new market options (Sablik, 2014). Tourist behaviors, including social behavior, are affected by the sharing economy (Cheng, 2016). This new form of consumption is considered communal bonding, which reduces the environmental effect of consumption because of consumer participation (Tussyadiah, 2016). Nica and Potcovaru (2015) found that this form of consumption is a way of sharing saved resources with others. Oskam and Boswijk (2016, p. 16) indicated the following three main factors when taking part in the sharing economy: 'convenience and price, the product or service itself, and word-of-mouth.'

First, price is identified as the major factor of Airbnb's appeal to tourists (Guttentag, 2015). Airbnb accommodations are often cheaper than traditional hotels. Airbnb is a win–win option for users who can find low-cost accommodations, and operators who can earn extra income (Dollberg, Shalev, & Chen, 2010; Ikkala & Lampinen, 2015; Oskam & Boswijk, 2016; Pizam, 2014). These accommodations offer discounts of up to 40% off the price of a hotel room (Yung, 2014). Many studies (Guttentag, 2015; Heo, 2016; Lehr, 2015; Nica & Potcovaru, 2015; OECD, 2016; Oskam & Boswijk, 2016; Tussyadiah, 2016; Weber, 2014; Zervas, Proserpio, & Byers, 2016) also found that Airbnb is cheaper than hotels, which is an attractive factor to young people, families, and people with low travel budgets (Varma, Jukic, Pestek, Shultz, & Nestorov, 2016). However, Poon and Huang (2017, p. 2438) found that Airbnb is 'less preferred when traveling with family,' which shows the need for further research from the perspective of family users.

Second, the product of Airbnb is distinct from the standardized hotel experience. Mody, Suess, and Lehto (2017) identified serendipity, localness, communitas, and personalization as the new dimensions of experience economy. The Airbnb platform emphasizes local advice and an authentic experience with Airbnb hosts. A unique local experience (Tussyadiah & Pesonen, 2016) and meaningful social encounter (Cheng, 2016) highlight the differences between Airbnb and traditional accommodation service. The current study argues that the different kinds of informal accommodations (staying with or without a host) and the benefit of receiving local advice and an authentic experience may not apply to all types of traveler needs.

Third, the real review system introduced word-of-mouth through this social platform. Varma et al. (2016) noted that Airbnb users are not loyal to Airbnb. Airbnb users will use the service in the future if they have satisfactory experiences, which demonstrate the importance of user reviews. This new consumption pattern offers increased value with low loyalty (Tussyadiah, 2016). Airbnb tourists/travelers tend to stay long and travel frequently, which can benefit the entire tourism industry (Cheng, 2016; Fang, Ye, & Law, 2016; Guttentag, 2015; Tussyadiah, 2016; Varma et al., 2016; Zervas et al., 2016). By contrast, family travelers have weak loyalty to a particular hotel brand (Choi et al., 2010). Thus, peer-to-peer reviews (P2P) offer valuable and informative data to users, which may be an ideal alternative for family travelers.

Travelers look for new and unique ways of traveling and sharing their experiences in social media (Yung, 2014). Mody et al. (2017) found that the positioning of the Airbnb platform is evidence of the experiential value of the sharing economy. They noted that experience is the key element of the hospitality and tourism industry. Social interaction between tourists and hosts may be considered a key factor that determines the perceived value, enjoyment (Tussyadiah, 2016), and satisfaction of tourists (Heo, 2016). Their social interaction with local hosts may be an important channel that promotes their experiences to others. This authentic and unique experience can also generate new tourism products and enhance destination attributes (Tussyadiah & Pesonen, 2016). By contrast, staying in hotels tends to offer a standard service format for family travelers. Existing studies in family travel and hotel selection demonstrate a number of key factors. Only a few recent studies have discussed how Airbnb offers value to young and family travelers. These studies did not particularly focus on family travelers. Therefore, the current study aims to explore whether Airbnb is a good choice for family travel by answering the following questions.

(1) What are the factors that influence the decision of booking Airbnb instead of a hotel?
(2) How does the Airbnb experience affect future travel patterns/behaviors?

Methodology

This study aims to investigate how the sharing economy shapes family travel in an accommodation service. This study adopts a qualitative approach given the limited understanding of family travelers' preference in Airbnb accommodation. This approach allows the researcher to investigate the phenomenon of Airbnb development by analyzing individual feelings toward the use of an alternative accommodation service for family travel. The interview strategy enables the interviewer to probe into the views and rationale of participants in choosing Airbnb for family travel. Semi-structured questions are designed for one-on-one interviews to understand the experience of participants of their stay at an Airbnb accommodation.

A purposive sampling technique was applied in this study, which was designed to elicit a small number of cases and provide the most information about family travelers using Airbnb. Sampling size is based on the judgment of the researcher and is normally composed of 30 cases or less (Teddlie & Yu, 2007). Participants were selected using snowball sampling. First, the researcher sent an interview invitation through a parents group in Hong Kong via the communication app WeChat. The group consists of 133 members who have children in primary and secondary schools. The researcher received nine potential parents who were willing to share their experiences. Eight interviews were successfully conducted through phone and face-to-face formats. Six referral cases in Mainland China and Australia were conducted through phone interviews, and two interviews from Taiwan were conducted by phone. All 16 one-on-one interviews were conducted in February 2017, which included phone and face-to-face interviews in Hong Kong, Mainland China, Taiwan, and Australia. Each interview lasted approximately 30 to 45 minutes using a digital recorder. Before the interview, each participant was briefed about the purpose of the study and confidentiality of personal data.

Participants consisted of 13 females and three males who have at least one child within the ages of 1 to 10 and who have experienced staying at an Airbnb within the last year. They used Airbnb at least once for their family travel within the past year. One participant is under 25 years old, four participants are 26 to 35 years old, seven participants are 36 to 45 years old, and four participants are 46 years old and above. Fifteen participants are currently based in Hong Kong, Mainland China, Taiwan, and Australia. Ten participants have a master's degree and higher, four participants have received a bachelor's degree, one an associate degree, and one has received a senior high school degree (Table 1).

All participants chose Airbnb for the travel purpose of family leisure. Most participants learned about Airbnb through friends and family recommendations, and a few learned about Airbnb through the internet while arranging accommodation. The average length of stay in Airbnb accommodation is 2.6 nights. Most participants stayed for at least once or twice a year, followed by three to four and five to six times. Only four participants experienced sharing accommodation with the host, whereas most stayed in independent flats, houses, or villas. All participants had experience of using the Airbnb service overseas. Among the 16 participants, only three had experience of using the domestic Airbnb service.

Table 1. Demographic information.

Participant	Gender	Age	Education Level	Number of child	Residence
1	M	46+	Bachelor	1	Hong Kong
2	F	26-35	Associate Degree	2	China
3	M	36-45	Senior High School	4	Taiwan
4	M	46+	Master and above	2	Taiwan
5	F	36-45	Master and above	2	Taiwan
6	F	36-45	Master and above	2	China
7	F	18-25	Master and above	2	Australia
8	F	36-45	Bachelor	1	Taiwan
9	F	46+	Master and above	2	Hong Kong
10	F	36-45	Master and above	2	Hong Kong
11	F	46+	Master and above	1	China
12	F	36-45	Master and above	1	China
13	F	26-35	Master and above	2	China
14	F	46+	Bachelor	2	China
15	F	26-35	Master and above	1	China
16	F	26-35	Bachelor	1	China

All interviews were transcribed verbatim and analyzed using NVivo 11. All data were coded into three major themes – namely, demographic information, reasons for choosing Airbnb, and experience of using Airbnb for family travel. First, interviewees were asked about their demographic information, which includes age, family members, occupation, education level, and income. They were then asked about their previous experiences using Airbnb, which include finding the Airbnb service, travel purpose, accommodation type, and length of stay. Second, they were asked about their reasons for choosing Airbnb instead of a hotel. Third, they were asked how the Airbnb experience differed from a hotel experience.

Findings and discussions of Airbnb experience

Interviewees were asked to share their first Airbnb experience, and most agreed that Airbnb has met or has exceeded their expectations. Some interviewees had unpleasant experiences, including differences in the photo uploaded on the website from the actual place and the difficulty of finding the place. Most unpleasant experiences are related to the former, such as old and dirty facilities (Interviewees 5, 7, 9, and 10). Interviewee 1 encountered difficulty finding the place and opening the door, which created an unpleasant experience. Not all Airbnb hosts would meet and open door for their guests. In fact, some interviewees did not get a chance to meet the hosts during their stay. Interviewee 14 had difficulty finding the place in a local community in France because she does not speak the language. Other participants could speak the local language, which made their travel easier. Therefore, language competency also affects the smooth travel of family members.

Interviewee 3 availed a secret accommodation in Bangkok, where the apartment complex does not welcome the Airbnb operation.

> Our first experience was an apartment in Bangkok. We were asked not to reveal the purpose of our stay. We could only mention to the security guard that we were there to meet friends. There was no contact with the host at all. We had to do it with all the instructions based on our previous communication. We felt like criminals staying in someone's house.

Staying in a city that does not welcome Airbnb can lead to an unsecure and unexpected experience. Two interviewees (3, 5) indicated that they may no longer use Airbnb for their family travel because of their unexpected and unsatisfying experiences.

The majority of interviewees who had a positive experience of using Airbnb consider the platform their first choice. This finding extends the view of Varma et al. (2016) who suggested that satisfied Airbnb users may possibly use the platform again in the future. This finding also reveals that dissatisfied users may no longer choose Airbnb for their future family trips, which involve children of various ages. Thus, each family prefers not to deal with unexpected situations. The current study finds that family travelers who are satisfied with Airbnb may likely use Airbnb for future family travels.

Selecting Airbnb

What are the factors that influence the decision to book Airbnb instead of a hotel? Most family travelers need more than one room. A hotel may only allow one child to stay in an existing bed, which may not fit a family with two kids, but paying for an additional room is costly. Interviewee 5 said: 'When you have to spend the same amount for motel or Airbnb, I would choose Airbnb.' Family travelers look for cost-efficient accommodation. Interviewees were asked of the percentage of hotel price that they would pay for Airbnb. Two-thirds of interviewees believed that price is the key indicator for their decision to choose Airbnb instead of a hotel. They agreed that Airbnb offers value for money. Thus, they can save money for other expenses. Six interviewees indicated that they would choose Airbnb if accommodation is 20% to 30% cheaper than hotel prices. Nine interviewees would consider Airbnb if the price is 30% to 50% lower than a hotel. Three interviewees do not take price as a key factor in their choice of accommodation, but they look for quality accommodation and unique experience for their stay. Most family travelers support the finding of existing studies, that price influences their choice of accommodation services (Guttentag, 2015; Oskam & Boswijk, 2016; Tsai et al., 2011; Wong & Chi-Yung, 2002; Yung, 2014). Other family travelers look for authentic and meaningful experiences rather than good price. The price of Airbnb is considered a key factor. This finding did not incorporate other factors in the following analysis.

Table 2 shows the factors that influence the decision of booking Airbnb rather than a hotel. A total of 222 text units were identified, which include special experience (120 text units, 46%), facilities and space (73 text units, 28%), and location (29 text units, 11%).

Special experience

Most interviewees used Airbnb for various reasons. 'Special experience' received the highest weight (120 text units, 46%) among all attributes. Special experience includes cultural exploration, local lifestyle, and interior and exterior of a house/building. Some interviewees purposely select a unique Airbnb accommodation to understand the local lifestyle and create a memorable experience (e.g. staying in a farmhouse, house of an artist, or a historical building).

> We love Japanese melon. We purposely went to a melon farm to purchase melons from the farmers. My son and I can speak Japanese. We interacted with local residents. I feel that we enjoyed our trip as more than just being tourists. (Interviewee 1)

The interviewees engaged in cultural exploration and lifestyle for themselves and their children. Interviewee 7 likes to stay in unique structures, such as a castle in Europe. Unlike staying in a standard hotel, they gained unique experiences by staying in a castle. These informal accommodations enriched their sense of belongingness in a home. Interviewee 4 mentioned that their family experienced rich culture when they stayed in a unique accommodation. Feelings toward home and cultural meanings are consistent with Obrador's (2012) finding on how families value these fundamental elements to enrich their well-being.

Interviewees 1 and 16 shared that they select their Airbnb accommodation to enrich their travel experience as a family. According to Interviewee 16, they stayed in an Airbnb for the cultural exploration of their children. They selected a farm house where they were required to take care of animals and the entire house. They needed to milk cows and water plants in the morning and make sure the gate is closed at night. Their family gained the real experience of living on a farm and claimed that the experience cannot be replicated by hotels. This finding supports the unique selling points of the

Table 2. Decisive factors.

Category and subcategories	Frequency of text units	Percentage	Salient points (key examples)
1.1 Special experience	120	46%	• We visited a farm in Japan and had a pleasant chat with a local farmer. It was a nice and special experience for us. (Interviewee 1) • The villa was near Disneyland. Thus, the theme of my children's rooms was Disney. They loved it! (Interviewee 6)
1.1.1 Esthetic	38	15%	• The decoration of the room is Japanese setting (tatami, wooden floor, and Japanese handicrafts). (Interviewee 10) • The landlords are art and travel lovers and have a lot of artworks from all parts of the world. (Interviewee 5)
1.1.2 Educational	35	13%	• My son could practice his Japanese with the Japanese landlord. (Interviewee 1) • In Australia, we stayed in a villa, which was more than 100 years old. The decoration and the facilities were historical and artistic. Thus, the children had great experience in this stay. (Interviewee 8)
1.1.3 Escapist	26	10%	• I cooked by myself, and all the vegetables were picked from the local farm. I have my own sheep and dog there. I felt like I was in my second home. (Interviewee 11) • I stayed in a local community and went to the local market by taking public transportation. I cooked by myself. I really enjoy the local lifestyle. (Interviewee 14)
1.1.4 Entertainment	21	8%	• We went to the Sapporo beer festival in Japan. The people there were nice. (Interviewee 1) • The night market in Thailand was nice. I love their food and handcrafts. (Interviewee 12)
1.2 Facility and space	73	28%	• The washing machine is important to our family when traveling in summer. (Interviewee 1) • We will choose Airbnb when traveling with many friends because it offers more space, and we feel at home. (Interviewee 12)
1.3 Location	29	11%	• Location is important and should be close to public transportation. (Interviewee 1) • Location is important when traveling with kids. The area should be safe and near public transportation. (Interviewee 10)
Total:	222	100%	

sharing economy by sharing saved resources (Nica & Potcovaru, 2015) as products or services (Oskam & Boswijk, 2016) to create a win–win situation for hosts and users (Dollberg et al., 2010; Ikkala & Lampinen, 2015; Oskam & Boswijk, 2016; Pizam, 2014). Moreover, involvement by participation enhances the travel experience of families and can be applied to any children-oriented accommodation service.

This study also finds that families treasure travel experiences with their children. Pine and Gilmore (1998) classified experience into four dimensions, namely, education, entertainment, escapist, and esthetic (Atwal & Williams, 2012). The finding of this study was also categorized into four elements, namely, experience economy, 1.1.1 esthetic (38 text units, 15%), 1.1.2 educational (35 text units, 13%), 1.1.3 escapist (26 text units, 10%), and 1.1.4 entertainment experience (21 text units, 8%).

The esthetic realm involves immersing themselves in a sensual environment (Quadri-Felitti & Fiore, 2012). Interviewees 12, 13, and 14 appreciated the distinctive culture of aesthetics of Airbnb accommodations. 'When I saw the collection of kitchenware, I felt that my host pays much attention to quality of life. I love to cook, and that immediately caught my eye' (Interviewee 15). Many interviewees discussed how they enjoyed the decoration and taste of their Airbnb host.

The realm of education is commonly applied in the tourism and hospitality industry to enrich the experience of travelers by offering informal and educational activities (Williams, 2006). Education is highly cherished by the interviewees who frequently mentioned that they treat their stays as educational experiences for their kids. Interviewee 14 mentioned that their interaction with the hosts, children, and pets expanded her son's horizon, and shared his reaction to a disabled dog adopted by the host. She believes that this educational experience cannot be gained from school. Certain interviewees also believe that allowing children to speak the local language and communicate with the host or local residents are ways of learning (Interviewees 1 and 10).

The escapist realm is a way to immerse themselves in a sensual environment (Quadri-Felitti & Fiore, 2012). Staying in someone else's house through Airbnb allows family travelers to immerse themselves in a new culture and lifestyle. Each escapist experience is distinct for the family. Many interviewees shared that they felt they were residents. 'I feel I am at home and part of the locals. The host has a farm where you are welcome to pick your cucumbers and tomatoes for salad' (Interviewee 11 in New Zealand).

The entertainment realm is a way to reflect engagement 'in a personal, memorable way' (Pine & Gilmore, 1998, p. 99). The interviewees enjoyed their stay because of the entertainment provided by local festivals, hosts, and surroundings. 'The Sapporo beer festival was being celebrated during our stay in Sapporo. We also joined the local festival where we received a warm welcome and had a great time' (Interviewee 1). Many family travelers opted to stay on a farm where they can have a completely different experience from their daily life.

> In Australia, the host has a glasshouse where he plants many vegetables. At night, we can see wild rabbits and kangaroos in the forest. I felt that we did not need to spend extra time and money to add another activity in our trip. (Interviewee 9)

These distinctive experiences highlight unique local and social interactions and differentiate Airbnb from hotel services, which supports the findings of Tussyadiah and Pesonen (2016). Family travelers can also experience the authentic features of staying in local houses by interacting with residents and hosts (Varma et al., 2016). Family travelers value learning opportunities for their children that schools do not offer.

Facility and space

Facility and space received the second highest weight (73 text units, 18%). This study identifies several key elements in the facilities category, which are highly relevant to the needs of traveling families (e.g. washing machine and kitchen). Kitchen is one of the key reasons for choosing an Airbnb accommodation. Participants agreed that a washing machine and kitchen are the key facilities they consider for choosing Airbnb instead of a standard hotel service. Children who travel outside their homeland may not eat local meals from restaurants. This phenomenon occurs when they travel to Western countries. Certain interviewees travel with groups of families and share Airbnb accommodations and cook while their kids play.

> My daughter was sick during our recent family trip. I was glad that I was able to cook something for her. When searching for accommodation service, I would look for kitchen facilities, which allow me to prepare simple food for my family. (Interviewee 5)

Interviewee 3 frequently travels with kids and parents who often prefer to eat simply in the Airbnb accommodation and save money. Thus, having a kitchen is highly important because family travelers can prepare comfort food and save on dining expenses.

Family travel involves at least two members in the trip. The 16 interviewees mentioned that their family travels involved three to ten members. A washing machine can reduce their packing and burden, and most interviewees mentioned the importance of having one. Children's clothes can easily get smelly and dirty when traveling the whole day, especially in the summer. Interviewee (1) stated: 'When staying in a hotel, you need to pay an unreasonable price for laundry service. Alternatively, if you are staying in an Airbnb, you can do it without any worry.' Most interviewees tend to find accommodation that provides child-friendly facilities, such as swimming pools, BBQ facilities, and playrooms. Airbnb offers a variety of accommodations that allow family travelers to select those that fit their needs. Washing machines and kitchen facilities are not mentioned in previous studies of family travel for selecting hotel criteria. Most hotels do not offer kitchen facilities and washing machines.

All interviewees indicated that they normally book at least two rooms. The cost of hotel rooms is usually high, and sometimes they cannot secure adjoining family rooms, which causes inconvenience. Many interviewees indicated that the size of hotel rooms is often limited, and children cannot run around or make much noise. Families in Airbnb accommodation do not have to worry about the noise. 'Most of the time, the hotel room is only around 40 m^2. However, the size of Airbnb accommodation is 400 m^2' (Interviewee 11). Other interviewees prefer renting a big house or villa, which they do not often experience, to enrich their holiday experience.

'We live in a very small apartment in Hong Kong. When we are traveling, we would choose a big house to satisfy our needs' (Interviewee 5). The benefit of staying in a house is having 'a living room where everyone can talk [communicate] and make the ambiance even better' (Interviewee 6). Facilities and space provide family travelers with more convenient support than hotels. Compared with Tsai et al.'s (2011) study about selecting hotel criteria, the current study finds that Chinese family travelers value facilities and space, which is similar to Western travelers who focus on room size. Hotels can offer good value for a laundry service for family travelers to reduce their burden.

Location

Location received the fifth highest weight among all attributes (27 text units). Most interviewees care about location for various reasons and purposes, such as being near tourist destinations, proximity to public transportation, supermarkets, shops, and restaurants, and a safe zone to stay in. Interviewees 1 and 14 mentioned that finding Airbnb accommodation creates problems, especially when traveling with kids. 'I would choose one near to railway station or a convenient transportation system' (Interviewee 1). Others do not consider location as a key factor. They can rent a car and find the place easily.

Family travelers who visit the city commonly look for convenient life function and location, where they can easily access public transportation. Those who visit the countryside commonly travel by car. Thus, the location may not be as important as other factors. Finding a safe area to stay is the key factor for them.

Word of mouth

Word of mouth received a total of 39 text units, which include 2.1.1 host, 2.1.2 safety, 2.1.3 cleanliness, 2.1.4 service, and facilities (Table 3).

Interviewees depend on recommendations by friends and online reviews when making their final decisions. Some interviewees experienced their first Airbnb through recommendations of friends. Interviewees 5, 7, and 11 strongly trust the recommendations and shared experiences of their friends. 'I introduced my Airbnb hosts from Canada and New Zealand to my friends and many of them also had a great time' (Interviewee 11). This outcome extends the findings of Kozak (2010) on the importance of word-of-mouth recommendations on hotel and Airbnb services. In terms of online

Table 3. Word of mouth.

Category and subcategories	Frequency of text units	Percentage	Salient points (key examples)
2.1 Word of mouth			• Reviews become the important things when I choose the place to stay. (Interviewee 15) • My friend suggested to use Airbnb because it offers much space and connecting rooms. I think the recommendation of friends is trustworthy. (Interviewee 7)
2.1.1 Host	13	33.4%	• I would read the online reviews first and choose the area that has good reviews. (Interviewee 15)
2.1.2 Safety	11	28.2%	• If the review says that the area is not secure, I will skip it. (Interviewee 8)
2.1.3 Cleanness	8	20.5%	• The most important factor is cleanliness. (Interviewee 2)
2.1.4 Service and facility	7	17.9%	• I saw on the online review that they provide pick-up service. Thus, I chose it. (Interviewee 2) • I care about the online review on room facilities. (Interviewee 5)
Total	39	100%	

Table 4. Legal issue.

Category and sub-categories	Frequency of text units	Percentage	Salient points (key examples)
3.1 Will not choose Airbnb if it is illegal	11	48%	• As a tourist, safety is very important for me. So if it is illegal, I will not choose it. (Interviewee 13)
3.2 Depends on regions and reviews	11	48%	• It depends on the region and also the reviews about the Airbnb lodging. For example, Japan, people there is very friendly, public security is also good and very safe! (Interviewee 1) • I will communicate with the landlord regarding some important and sensitive issues first and then decide whether to choose it or not. (Interviewee 14)
3.3 Still chooses Airbnb, if it is illegal	1	4%	• I will still choose it because the price is very low and attractive. (Interviewee 4)
Total:	23	100%	

review, interviewees first review host profiles, pictures of the place, and user comments. 'My first Airbnb experience was not as good as I had expected mainly because I did not pay much attention to the reviews of users on the host' (Interviewee 9). Interviewees 6 and 13 pay attention to host and family reviews and attitude toward children. This outcome supports the study of Ert, Fleischer, and Magen (2016) on how host profiles influence the decision-making of users.

Safety is also a crucial consideration for family travelers, which consists of area, accommodation, and host. 'When I was looking for a place to stay in New York City, I could not find a reasonable accommodation. Then I found one in Brooklyn. I checked the review about safety in the area and the apartment itself' (Interviewee 6). Interviewee 2 believes that most guests leave moderate comments about their experience. However, this interviewee also stated: 'Sometimes you can spot some key points if you read carefully. The host has a very cute dog, but dog hair is everywhere. I would try to avoid this accommodation.' An unfriendly host is the second keyword that should be avoided. Certain interviewees mentioned that unfriendly hosts can be related to unreasonable charges. Interviewee 7 was asked to pay an extra deposit when her family arrived at the destination, which created a negative perception of their travel experience.

Service and facility were the factors that were least considered in reviews. Service includes airport pick-up, local tour service, and meals. Interviewee 11 stays with the same host in New Zealand every year. Thus, she feels that she keeps going back to a second home in New Zealand. 'My host came to pick us up at the airport and assisted us to book a South Island tour.' Most interviewees noted that facilities and space are essential to their choice of stay. Interviewee 14 often travels with a pet and has difficulty finding a hotel to host them. Other Airbnb hosts accommodate pets, which made her family travel easy.

Legal and safety issues

Safety is a key element from the perspective of family travelers. The interviewees shared their views about the legal issues in Airbnb operations, which most interviewees were not aware of. Interviewees 4 and 6 know that Airbnb in some cities, such as San Francisco and Amsterdam, are required to pay taxes. All interviewees indicated that Airbnb should be legalized, which can benefit users, hosts, and the local tourism industry. Interviewee 16 mentioned the need for a certain type of insurance implementation, which can protect users and hosts. Approximately half of the interviewees stated that they will not choose Airbnb if they know its operation is illegal. However, half of the interviewees stated that they may still select Airbnb if a city has a good reputation based on safety and online reviews. Interviewee 1 mentioned that Japan has a good reputation and he does not mind using Airbnb in this country even if Airbnb is not regulated. This outcome may relate to the previous positive experiences of interviewees who have stayed in an Airbnb accommodation. This outcome also reflects how family travelers value online reviews. The interviewees appreciate the safety of Airbnb and its regulated accommodation service for family travel. Governments and local authorities must handle the development of Airbnb.

Most interviewees believe that Airbnb development influences the patterns of family travel because they feel they are going to a second home in another city or country. Interviewee 10 stated: 'I used to plan our family trip to visit different places fully. Now, we slow down our plan and also take local host advice that enriches our travel experience.' Other participants also took local host advice and ended up having an unexpected experience. Interviewees 1 and 13 mentioned: 'We feel we are part of the locals. Just like going to the market and buying some vegetables and meat for dinner.' Interviewees 2, 4, and 15 shared that they do not plan their trips in detail. Other participants mentioned that their travel pattern changed to 'pack and go' from making a plan first. This outcome supports the commercial of Airbnb, of 'welcome home,' which makes users feel like everywhere is their home. They simply fly to a city to experience local authenticity and visit well-known sites.

Not all family members enjoy staying in an Airbnb accommodation. Interviewee 14 mentioned that her husband prefers to stay in a hotel where he can expect standard service and facilities. However, she plans to bring her kids to a boat house in Hong Kong with her friends. This finding demonstrates a unique experience of using Airbnb as something that is highly welcome by some family travelers compared with a standardized service and setting. Airbnb gradually changes the travel patterns of families and closes the distance between travelers and local hosts. Family travelers value their local experience and alternative accommodation choice. This outcome supports the finding of Varma et al. (2016) that Airbnb is suitable for family travel because it provides alternative accommodation services; however, the findings do not support Poon and Huang (2017). In addition, users who have positive experiences with Airbnb tend to use this service in the future, whereas users with negative experiences prefer not to select it again. However, hotel cost is sometimes still higher than Airbnb, which makes users take the risk again.

Conclusion

This study explores the key factors for families to choose the Airbnb accommodation service. Previous studies have discussed how children affect family decision-making for holidays, but do not include this new alternative accommodation service, Airbnb. Price is often a major consideration for family travelers in selecting their accommodation service, whereas others chose to stay in Airbnb for a unique experience. This study demonstrates that special experience is the most important factor, as it is a way of learning about a new culture for family members, followed by facilities and space, online reviews, location, and recommendations of friends.

Family travelers look for more than a hotel room and standard service. Airbnb can offer this service and other unique experiences, which cannot be duplicated by standard hotel accommodation services. For family travelers, staying in an Airbnb accommodation 'corresponds to customer participation and describes the connection, or environmental relationship' within their stay (Pine & Gilmore, 1998, p. 101). This study finds that family travelers value the esthetic realm the most within their experience followed by education, escapism, and entertainment realms. This outcome can be further examined by a quantitative approach in the future.

Airbnb offers flexible facilities and space, which can fulfill the needs of families; these facilities include a kitchen, washing machine, and additional space. Online reviews are also highly valued by family travelers, which indicate that Airbnb hosts should maintain positive reviews, especially regarding their image and presentation, accommodation safety, cleanliness, service, and facilities. Airbnb hosts can arrange activities to engage family members and enrich their holiday experience. Good quality accommodation can attract additional family travelers rather than low price. Location is not an important factor among the attributes as long as hosts provide clear directions to reach the accommodation location. Family travelers who use Airbnb slightly differ from family travelers who use resort hotels (Kang et al., 2003; Khoo-Lattimore et al., 2015).

Family travelers believe that Airbnb should be regulated and supervised by local authorities to ensure the quality of the tourism industry. The current study extends existing studies of Airbnb operation from the perspectives of family travelers. The outcome supports the finding that Airbnb accommodation is an attractive accommodation service for family travelers (Varma et al., 2016), which contradicts Poon and Huang (2017).

This study identifies the key factors of how family travelers select the Airbnb service with regard to their accommodation. This finding is a practical contribution to hotels and Airbnb hosts. Hotel and Airbnb operators can provide family travelers with additional suitable facilities and experiences by understanding their needs. Family travelers look for a special participation experience that can enrich their travel. Besides offering good hygiene standards, this study suggests that Airbnb hosts, who have abundant natural resources and access to farms and gardens, should organize feasible programs for children, to enrich their stay. Certain interviewees found time to read reviews based on existing views of those using a P2P service. Despite the current smart surface of the Airbnb platform, this study suggests organizing a special group of children-friendly hosts and services, which can easily guide family users to adopt this informal accommodation service. Airbnb is related to the trend of the sharing economy that offers an alternative accommodation service for family travel, which creates a new trend regarding travel behavior and which influences hoteliers to rethink the promotion of a children-friendly stay and experience.

In terms of the legal issues of the Airbnb operation, the majority of participants do not know whether Airbnb is regulated by local authorities or not. This finding creates a concern for family travelers, especially when they travel with young children. Interestingly, certain participants use Airbnb based on online reviews and where they are traveling to. In certain cities, licensed guest house hosts also use Airbnb as a distribution channel to promote their business. This licensed guest house can also be a way of ensuring the quality and standards to users. Online reviews is still a key indicator to travelers in that they are closely related to guest experiences. Providing a pleasant experience is also a task that Airbnb hosts should learn and offer.

Limitation and future study direction

This study explores the factors that influence family travelers to select Airbnb as an alternative accommodation service. This study encounters unavoidable challenges, which can serve for future research opportunities and directions. First, this study applies one-on-one interviews to provide information filtered through the views of participants. Each participant experienced using the Airbnb accommodation service at least once. During the interview process, participants shared their experiences of selecting Airbnb, which may also reflect their overall experience. Interpreting the views of participants may result in biased responses and not all participants are articulate and perceptive (Creswell, 2009). All participants are of Chinese background. Their views can represent the Chinese family experience. The findings of this study set the foundations for research on family travelers who use Airbnb, which can be extended to different approaches in order to investigate the needs and experiences of family users. Further research recommendations include: (1) applying a quantitative method to guide the needs and experiences of family travelers in the Airbnb service, and (2) replicating the study in different cultures.

Disclosure statement

No potential conflict of interest was reported by the author.

Funding

This paper and research project (Project Number: 2015.A6.066.16B) was funded by the Public Policy Research Funding Scheme from Policy Innovation and Co-ordination Office of the Hong Kong Special Administrative Region Government.

ORCID

Pearl M. C. Lin http://orcid.org/0000-0002-4320-3866

References

Atwal, G., & Williams, A. (2012). Is this Shangri-La? The case for authenticity in the Chinese and Indian hospitality industry. *Journal of Brand Management, 19*(5), 405–413.

Belch, G. E., Belch, M. A., & Geresino, G. (1985). Parental and teenager child influences in family decision-making. *Journal of Business Research, 13*(2), 163–176.

Berey, L. A., & Pollay, R. W. (1968). The influencing role of the child in family decision making. *Journal of Marketing Research, 5*(1), 70–72.

Cheng, M. (2016). Sharing economy: A review and agenda for future research. *International Journal of Hospitality Management, 57*, 60–70.

Choi, H. Y., Lehto, X., & Brey, E. T. (2010). Investigating resort loyalty: Impacts of the family life cycle. *Journal of Hospitality Marketing & Management, 20*(1), 121–141.

Creswell, J. W. (2009). *Research design: Qualitative, quantitative, and mixed methods approaches* (3rd ed.). Thousand Oaks, CA: Sage Publications.

Decrop, A. (2005). Group processes in vacation decision-making. *Journal of Travel & Tourism Marketing, 18*(3), 23–36.

Dollberg, D., Shalev, O., & Chen, P. (2010). 'Someone's been sleeping in my bed!' Parental satisfaction associated with solitary and parent–Child co-sleeping in Israeli families with young children. *Early Child Development and Care, 180*(7), 869–878.

Ert, E., Fleischer, A., & Magen, N. (2016). Trust and reputation in the sharing economy: The role of personal photos in Airbnb. *Tourism Management, 55*, 62–73.

Fang, B., Ye, Q., & Law, R. (2016). Effect of sharing economy on tourism industry employment. *Annals of Tourism Research, 57*, 264–267.

Flurry, L. A. (2007). Children's influence in family decision-making: Examining the impact of the changing American family. *Journal of Business Research, 60*(4), 322–330.

Flurry, L. A., & Burns, A. C. (2005). Children's influence in purchase decisions: A social power theory approach. *Journal of Business Research, 58*(5), 593–601.

Gram, M. (2005). Family holidays. A qualitative analysis of family holiday experiences. *Scandinavian Journal of Hospitality and Tourism, 5*(1), 2–22.

Guttentag, D. (2015). Airbnb: Disruptive innovation and the rise of an informal tourism accommodation sector. *Current Issues in Tourism, 18*(12), 1192–1217.

Heo, C. Y. (2016). Sharing economy and prospects in tourism research. *Annals of Tourism Research, 58*, 166–170.

Ikkala, T., & Lampinen, A. (2015). *Monetizing network hospitality: Hospitality and sociability in the context of Airbnb.* Paper presented at the Proceedings of the 18th ACM Conference on Computer Supported Cooperative Work & Social Computing, Vancouver, BC, Canada.

Kang, S. K., Hsu, C. H. C., & Wolfe, K. (2003). Family traveler segmentation by vacation decision-making patterns. *Journal of Hospitality & Tourism Research, 27*(4), 448–469.

Khoo-Lattimore, C., Prayag, G., & Cheah, B. L. (2015). Kids on board: Exploring the choice process and vacation needs of Asian parents with young children in resort hotels. *Journal of Hospitality Marketing & Management, 24*(5), 511–531.

Kim, S. S., Choi, S., Agrusa, J., Wang, K.-C., & Kim, Y. (2010). The role of family decision makers in festival tourism. *International Journal of Hospitality Management, 29*(2), 308–318.

Kozak, M. (2010). Holiday taking decisions – The role of spouses. *Tourism Management, 31*(4), 489–494.

Lawson, R. (1991). Patterns of tourist expenditure and types of vacation across the family life cycle. *Journal of Travel Research, 29*(4), 12–18.

Lehr, D. D. (2015). *An analysis of the changing competitive landscape in the hotel industry regarding Airbnb* (Master's Thesis). Dominican University of California.

Lehto, X. Y., Choi, S., Lin, Y. C., & MacDermid, S. M. (2009). Vacation and family functioning. *Annals of Tourism Research, 36*(3), 459–479.

Lehto, X. Y., Fu, X., Li, H., & Zhou, L. (2017). Vacation benefits and activities: Understanding Chinese family travelers. *Journal of Hospitality & Tourism Research, 41*(3), 301–328.

Lehto, X. Y., Lin, Y.-C., Chen, Y., & Choi, S. (2012). Family vacation activities and family cohesion. *Journal of Travel & Tourism Marketing, 29*(8), 835–850.

Lin, P. M. C., Tung, V. W. S., Qiu Zhang, H., & Gu, Q. (2018). Tourist experience on memorable hospitality services. *Journal of China Tourism Research, 14*(2), 123–145.

Mody, M. A., Suess, C., & Lehto, X. (2017). The accommodation experiencescape: A comparative assessment of hotels and Airbnb. *International Journal of Contemporary Hospitality Management, 29*(9), 2377–2404.

Nanda, D., Hu, C., & Bai, B. (2007). Exploring family roles in purchasing decisions during vacation planning: Review and discussions for future research. *Journal of Travel & Tourism Marketing, 20*(3–4), 107–125.

Nica, E., & Potcovaru, A.-M. (2015). The social sustainability of the sharing economy. *Economics, Management, and Financial Markets, 10*(4), 69–75.

Nickerson, N. P., & Jurowski, C. (2001). The influence of children on vacation travel patterns. *Journal of Vacation Marketing, 7*(1), 19–30.

Obrador, P. (2012). The place of the family in tourism research: Domesticity and thick sociality by the pool. *Annals of Tourism Research, 39*(1), 401–420.

OECD. (2016). *Policies for the tourism sharing economy.* Paris: Author.

Oskam, J., & Boswijk, A. (2016). Airbnb: The future of networked hospitality businesses. *Journal of Tourism Futures, 2*(1), 22–42.

Panja, T., & Millard, P. (2014, April 25). *Some world cup fans will really be slumming.* Retrieved from http://www.bloomberg.com/news/articles/2014-04-24/some-of-brazils-world-cup-visitors-will-stay-in-slums

Park, O. -. J., Lehto, X., & Park, J.-K. (2008). Service failures and complaints in the family travel market: A justice dimension approach. *Journal of Services Marketing*, *22*(7), 520–532.
Pine, B. J., & Gilmore, J. H. (1998). The experience economy. *Harvard Business Review*, *76*(6), 99–101.
Pizam, A. (2014). Peer-to-peer travel: Blessing or blight? *International Journal of Hospitality Management*, *38*, 118–119.
Poon, K. Y., & Huang, W.-J. (2017). Past experience, traveler personality and tripographics on intention to use Airbnb. *International Journal of Contemporary Hospitality Management*, *29* (9), 2425–2443.
Quadri-Felitti, D., & Fiore, A. M. (2012). Experience economy constructs as a framework for understanding wine tourism. *Journal of Vacation Marketing*, *18*(1), 3–15.
Roblek, V., Stok, Z. M., & Mesko, M. (2016). *Complexity of a sharing economy for tourism and hospitality*. Paper presented at the Faculty of Tourism and Hospitality Management in Opatija. Biennial International Congress, Tourism & Hospitality Industry, Opatija, Croatia.
Rothenfluh, F., Germeni, E., & Schulz, P. J. (2016). Consumer decision-making based on review websites: Are there differences between choosing a hotel and choosing a physician? *Journal of Medical Internet Research*, *18*(6), e129.
Sablik, T. (2014). The sharing economy: Are new online markets creating economic value or threatening consumer safety? *Econ Focus(4Q)*, 12–15.
Teddlie, C., & Yu, F. (2007). Mixed methods sampling: A typology with examples. *Journal of Mixed Methods Research*, *1*(1), 77–100.
Thornton, P. R., Shaw, G., & Williams, A. M. (1997). Tourist group holiday decision-making and behaviour: The influence of children. *Tourism Management*, *18*(5), 287–298.
Tsai, H., Yeung, S., & Yim, P. H. (2011). Hotel selection criteria used by mainland Chinese and foreign individual travelers to Hong Kong. *International Journal of Hospitality & Tourism Administration*, *12*(3), 252–267.
Tussyadiah, I. P. (2016). Factors of satisfaction and intention to use peer-to-peer accommodation. *International Journal of Hospitality Management*, *55*, 70–80.
Tussyadiah, I. P., & Pesonen, J. (2016). Impacts of peer-to-peer accommodation use on travel patterns. *Journal of Travel Research*, *55*(8), 1022–1040.
Varma, A., Jukic, N., Pestek, A., Shultz, C. J., & Nestorov, S. (2016). Airbnb: Exciting innovation or passing fad? *Tourism Management Perspectives*, *20*, 228–237.
Wang, K. C., Hsieh, A. T., Yeh, Y. C., & Tsai, C. W. (2004). Who is the decision-maker: The parents or the child in group package tours? *Tourism Management*, *25*(2), 183–194.
Weber, T. A. (2014). Intermediation in a sharing economy: Insurance, moral hazard, and rent extraction. *Journal of Management Information Systems*, *31*(3), 35–71.
Williams, A. (2006). Tourism and hospitality marketing: Fantasy, feeling and fun. *International Journal of Contemporary Hospitality Management*, *18*(6), 482–495.
Wong, A., Ap, J., & Li, L. (2001). *Family travel and perceptions on lodging facilities*. Paper presented at the Proceedings of Asia Pacific Tourism Association 7th Annual Conference, Philippines.
Wong, K. K. F., & Chi-Yung, L. (2002). Predicting hotel choice decisions and segmenting hotel consumers: A comparative assessment of a recent consumer based approach. *Journal of Travel & Tourism Marketing*, *11*(1), 17–33.
Yung, V. (2014). My Airbnb year in Hong Kong: 'Big fat American' discovers hidden sides to the city. *South China Post*. Retrieved from http://www.scmp.com/print/lifestyle/travel/article/1656453/expat-explores-hong-kong-year-short-term-stays-through-airbnb
Zervas, G., Proserpio, D., & Byers, J. (2016). *The rise of the sharing economy: Estimating the impact of Airbnb on the hotel industry*. Boston U. School of Management Research Paper No. 2013–16.

Index

Note: Figures are indicated by *italics*. Tables are indicated by **bold**.